Sustaining High Performing Public Enterprises: Case Study of National Water and Sewerage Corporation, Uganda

Sustaining High Performing Public Enterprises: Case Study of National Water and Sewerage Corporation, Uganda

Silver Mugisha

IWA PUBLISHING

Published by **IWA Publishing**
Alliance House
12 Caxton Street
London SW1H 0QS, UK
Telephone: +44 (0)20 7654 5500
Fax: +44 (0)20 7654 5555
Email: publications@iwap.co.uk
Web: www.iwapublishing.com

First published 2019

British Library Cataloguing in Publication Data
A CIP catalogue record for this book is available from the British Library

ISBN: 9781789060249 (print)
ISBN: 9781789060256 (eBook)
ISBN: 9781789060263 (ePUB)

Contents

List of Acronyms

AFD	French Development Agency
AfDB	African Development Bank
AfWA	African Water Association
APA	Arrears Payment Agreements
APC	Area Performance Contract
BSC	Balanced Score Card
CBO	Community-based Organisation
CDC	Centre for Disease Control
CP	Communications and Public Relations
CRM	Customer Relations Module
CRU	Corporate Relations Unit
CSI	Customer Satisfaction Index
CUG	Common User Group
DSN	Daily Staff Newsletter
EMM	Executive Management Meeting
ESD	External Services Department
FoT	Faculty of Technology
IDAMCs	Internally Delegated Area Management Contracts
IDES	Institutional Development and External Services
IHE-Delft	Institute of Hydraulic Infrastructure, Delft
IPAs	Individual Performance Agreements
IREC	International Resources Centre
ISDPs	Infrastructure Service Delivery Plans
IWA	International Water Association
KCCA	Kampala Capital City Authority
KFW	German Development Bank

KWSA	Kampala Water Supply Area
KYC	Know Your Customer
LK	Leo Kibirango
MD	Managing Director
M&E	Monitoring and Evaluation
NEMA	National Environment Management Authority
NFA	National Forestry Authority
NGO	Non-Governmental Organisation
NPM	New Public Management
NRW	Non-Revenue Water
NWSC	National Water and Sewerage Corporation
OIF	Operational Investment Fund
OMGs	One Minute Goals
PACE	Performance Autonomy and Creativity Enhancement
PhD	Doctor of Philosophy
PPDA	Public Procurement and Disposal Authority
PSP	Public Stand Posts
PSSS	Performance-based Salary Scale Structure
PURC	Public Utilities Research Center
QAC	Quality Assurance Committee
RPMs	Resident Project Manager
SCAP100	100% Service Coverage Acceleration Programme
SD	Strategic Direction
SPAs	Strategic Priority Areas
SSP	Sky Performance Programme
SUM	Sector Utility Management
SWALOP	Service Pipes Water Loss Prevention
SWAS	School Water and Sanitation
TAPs	Tactical Action Plans
UCE	Uganda Certificate of Education
UK	United Kingdom
UNBS	Uganda National Bureau of Standards
UNESCO	United Nations Educational, Scientific and Cultural Organisation
USA	United States of America
WACOCO	Water Community Communications Clubs
WALOPU	Water Loss Prevention Unit
WSSP	Water Supply Stabilisation Programme
YWPs	Young Water Professionals
WSS	Water and Sanitation Services
ENR	Environment and Natural Resources
DWD	Directorate of Water Development
DWRM	Directorate of Water Resources Management

DEA	Directorate of Environmental Affairs
TSU	Technical Support Unit
WSDF	Water Supply Development Facility
WMZ	Water Management Zone
UNMA	Uganda National Meteorology Authority
UWASNET	Uganda Water and Sanitation Network

About the Author

Silver Mugisha holds a PhD in Engineering and Economics from Makerere University, Uganda. His PhD research, which was carried out in collaboration with the Public Utilities Research Center at the University of Florida, USA, focused on areas of performance monitoring, incentive design and productivity analysis. He has published a book, *Utility Benchmarking and Regulation in Developing Countries: Practical Application of Performance Monitoring and Incentives* and over twenty policy and research papers.

Dr Mugisha has been the Managing Director of the National Water and Sewerage Corporation (NWSC) since August 2013. He has worked, on external consultancies, in Uganda and other countries. He was Vice-President of the International Water Association (IWA) in 2016–2018 and Vice-President of the African Water Association (AfWA) for the East African Region in 2018–2020. Dr Mugisha is a Senior Research Fellow at the Public Utilities Research Center (PURC), University of Florida, USA. He is also a fellow of the International Water Association and of the SENSE Research School of the Netherlands.

In addition, Dr Mugisha has attended a number of certificate courses from different institutions. These include, among others, certificates in Utility Regulation and Strategy from PURC, University of Florida, USA (2002); Private Sector Participation and Financing from Technical University of Berlin, Germany (2002); Result Oriented Change Management from African Management Development Centre, Tanzania (2002); Private Sector Participation from Stone and Webster Consultants and Catalyst 2, UK (2004). Other certificates gained include: Corporate Governance and Enhancing Board Effectiveness from Institute of Corporate Governance Uganda (2007); Public-Private Partnerships and Contractualisation in Water Supply Systems from World Bank Institute, USA/ AfWA (2008); Corporate Governance, Leadership, Stakeholder Management, Strategic Human Resource Management, Performance Management, and Emotional Intelligence from LK Consultancy, Kampala, Uganda (2016); Enterprise Risk Management, Risk Oversight and Cyber Resilience from BMK Advanced Corporate Training and Skills Development Centre, Johannesburg, RSA (2016); Role and Functions of Board and Management from LK Consultancy, Kampala, Uganda (2017); Career Development and Succession Planning from Meirc Training and Consulting, Dubai, United Arab Emirates (2017).

Dr Mugisha finds time to carry out social work in communities, religious institutions and schools. He has passion for contributing to the career growth of young people, engaging them in coaching and mentoring programmes. He is a strong advocate of strong attitudinal competences on top of knowledge and skills development for the upcoming and promising generation. He also likes motivating children to engage in climate-change remediation measures, environmental issues, hygiene management and efficient water management practices.

Preface

Professor Sanford (Sandy) Berg

I was honoured to be asked to provide an introductory preface to this volume. I have co-authored a number of articles with Silver Mugisha, so I have become familiar with the NWSC story over the years. However, this volume brings in brand new material on the strategies adopted and performance levels achieved over the past half-decade – a time of dramatic change and new responsibilities for the organization. It is remarkable that the achievements of the Ugandan water and sanitation utility have been sustained for two decades. Silver, the new Managing Director has maintained and enhanced the collaborative and innovative efforts of William Muhairwe (who had laid the foundations for further growth through his inspirational early leadership). The staff talents developed in those early years have continued to be nurtured and strengthened under Silver's leadership. This book identifies specific steps taken by NWSC to maintain that early momentum. Exhibits (statements) by a number of managers underscore the professionalism and commitment of key individuals who make good things happen!

The expansion in the size and scope of NWSC (with 11 towns in 1998, 23 in 2011, 141 in 215, and 218 in 2017) suggests a dramatic re-orientation of the company in recent years – from one focused on larger cities to a truly *national* utility that is addressing the needs of much smaller towns throughout Uganda. There is no doubt that this new vision is one that national political leaders sought and that was ultimately adopted and implemented by the leadership within

NWSC. The challenges associated with such an expansion cannot be underestimated – given the resource constraints facing the organization. Yet its professionals have taken on these new responsibilities and devised mechanisms for coping with smaller entities – applying (and adapting) lessons learned in the first decade of NWSC's process of reform and revitalization.

Silver and I wrote an article[1] on leadership that is relevant here: 'First, the terms *leader* and *leadership* need definitions. We can say that a leader provides direction when the right direction is already known. According to Heifetz (1994) in Jamison and Castaneda (2011), "Leadership mobilizes people to tackle difficult and often ambiguous problems and circumstances." That is, when the challenges and potential policies are not fully understood, leadership is required. Heifetz goes on to differentiate between *authority work* and *leadership*. According to Heifetz, "Authority work (1) provides solutions by applying established procedures; (2) protects people and the "system" from external threats, providing stability to the organization; (3) restores order so that work can continue when internal or external forces disrupt the normal performance of work; and (4) maintains norms so people can work by following established rules and procedures. In contrast, leadership work (1) identifies challenges by questioning how problems should be defined and pointing out why technical solutions cannot solve adaptive problems; (2) discloses threats by calling attention to fundamental changes in external forces that threaten the status quo and defy traditional solutions; (3) exposes real conflicts or facilitates their emergence so that those involved face and work through tough choices; and (4) challenges norms to ensure that "solutions" are not adopted before the new environment is fully understood, the real conflicts are resolved, and key trade-offs are made." '

Leadership requires that managers 'stir and steer'. Adaptive leadership requires the stakeholder setting to be stirred in order to surface problems, contradictions, and opportunities. So technically sound work on operations and cash flows is just the starting point. Performance results need to be communicated to key constituencies, including politicians. As we wrote earlier, '… the system also needs to be steered, not in the sense of leading a particular direction, but rather ensuring learning, providing opportunities for resolving conflict, and orchestrating experiments that lead to "next practices".' NWSC has been remarkably successful in this process precisely because challenges were identified, legitimate conflicts were resolved, and issues were addressed by capable and highly motivated individuals.

---

[1]Mugisha S. and Berg S. V. (2017). Adaptive Leadership in Water Utility Operations: the Case of Uganda, *Sustainable Water Resources Management*, **3**(2), 171–179.

Citing:

Heifetz R. A. (1994). Leadership Without Easy Answers. Harvard University Press, Cambridge, USA.

Jamison M. A. and Castaneda A. (2011). Reset for regulation and utilities: leadership for a time of constant change. *Electricity Journal*, **24**(4), 86–93.

This book outlines how NWSC leadership developed a new strategy that has made it effective even as its responsibilities grew exponentially. From the internal standpoint, managers gave attention to *operational excellence* (addressing water losses and water quality among other issues), process innovation (with an emphasis on universal service, information technology systems and risk management), and strategic alliances (with schools, young professionals, environmentalists, and other stakeholders). Focusing on *people*, NWSC emphasized enhancing employee productivity via incentives and other motivating factors, including national and international awards. *Financial sustainability* was supported through cost-containment and revenue-enhancing activities, while recognizing the political need for a pro-poor approach to pricing. Finally, the book examines NWSC's outward gaze – towards *customers and wider civil society*.

The concluding paragraph of our joint article on adaptive leadership seems appropriate here: '… without a willingness of leaders to go beyond a technical (engineering) orientation to one that seeks to engage external stakeholders and address emerging challenges, NWSC would not be winning awards. It might be easier to defer addressing genuine conflicts such as the one between those who would focus on a few cities and those who seek a wider impact for the utility. However, confronting and resolving differences ultimately improves the financial and political sustainability of the organization. Nothing is risk-free – moving into uncharted terrain presents more challenges than maintaining the status quo. Nevertheless, the case of NWSC suggests that adapting to changing expectations and adopting new tactics can lead to greater legitimacy for the utility and significant improvements in sector performance.'

NWSC's success was not accidental – it has been successful by *design*, through strategies devised collaboratively. Talented managers have recognized that they needed to become politically aware, communicating with politicians about the accomplishments and goals of the utility – becoming somewhat insulated from excessive external interference. The managerial leadership has sought data-driven decisions – requiring the collection and authentication of information. We know that decision-makers manage what they measure: the organizational culture of NWSC highlights measurement and transparency. Strong internal incentives direct managerial attention to issues that affect operations, employees, finances, and external stakeholders. Throughout the process, as Silver emphasizes in this volume, NWSC has supported capacity building. The organization's record of staff publications and advanced professional certifications is remarkable.

Let me share seven elements that (in my opinion) have affected NWSC's organizational performance:

Information – *information matters*: the collection and authentication of data is necessary to identify trends, understand current patterns of performance, and determine realistic targets for utilities; technical skills and on-going capacity building support such initiatives. I would assert that NWSC scores at the top of the scale on this element.

Institutions – *social structures and organizational cultures matter*: the former reflect cultural norms and customs. Cultural traditions, the rule of law, and political arrangements are the foundations of infrastructure systems. These are beyond the control of NWSC managers, but can affect operations to a significant extent. For example, the laws that establish roles and responsibilities for stakeholders constrain the various actors in terms of how they interact with utility operators. NWSC has utilized a variety of communications channels to relate to schools and other components of civil society.

Interests – *special interests matter:* every stakeholder benefits in different ways from managerial decisions. The utility itself is part of a system, which includes the legislature, government agencies, courts, regulators, and unions. Thus, inter-organizational collaboration is essential for improved sector performance. Communication is central to high-performing companies. NWSC has emphasized improving its relationship with external interests.

Incentives – *incentives matter:* decision-makers behave in accordance with payoffs associated with different outcomes; every managerial rule rewards or penalizes actions affecting utility performance. There is no doubt that NWSC has taken the use of incentives seriously – applying them at all stages of production.

Ideas – *ideas matter*: each of us brings a conceptual framework to our decisions ('believing is seeing'); new perspectives can serve as catalysts for activities that improve the operation and financial sustainability of utility operators. NWSC both applies and generates ideas that improve operations and allow staff to achieve their full potential.

Ideals – *values matter*: when we are clear about our objectives and communicate those priorities to stakeholders, the resulting dialogue can clarify our goals and promote greater consensus regarding organizational objectives. Ideologies, on the other hand, tend to focus on specific policies – which limits discussion. NWSC has clear priorities and bases incentives on quantified and authenticated outcomes.

Individuals – *people matter*: leadership is essential for improved sector performance; no matter how dysfunctional or inefficient current arrangements are, someone is benefiting from them – which implies that overcoming institutional inertia requires strong leadership.

I would argue that while each of the above elements influences performance, individuals are the most important determinants of organizational performance. The NWSC experience shows how a company has been able to recruit, retain, and re-tool its talented employees. Policies are not self-implementing. Leadership matters – at all levels of an organization. That is the key lesson from this volume.

Sanford Berg
Professor Emeritus, Economics
Senior Fellow, Public Utility Research Center
University of Florida, USA

Acknowledgements

A book volume like this cannot be successfully written without assistance of some friends and supporting environment. First of all, glory be to God who always gives us life and superintends over good efforts in making things happen for the benefit of His creation. I would like to thank the Government of the Republic of Uganda under H.E. President Y.K. Museveni for putting in place a conducive institutional, legislative and policy environment that has led to successful evolution of a case study of the National Water and Sewerage Corporation (NWSC) for over fifteen years. Without a good operating environment, the specific internal reforms that have taken place in NWSC would not have materialised. Special thanks go to Hon. Prof. Ephraim Kamuntu, the late Hon. Maria Mutagamba, Hon. Cheptoris, Hon. Ronald Kibule, and Hon. Dr Kitutu whose political leadership of NWSC's parent Ministry of Water and Environment has provided the requisite good policy framework. I also give special thanks to the Ministry's technical team led by former and current Permanent Secretaries Mr David O.O. Obong and Mr Alfred Okot Okidi, respectively, for the support and oversight.

I extend special thanks to the former Boards and current Board of NWSC led by Eng. Dr Christopher Ebal for the excellent strategic guidance and support. Special gratitude goes to my predecessors: Hon. Eng. Hilary Onek and Dr William T. Muhairwe who laid a strong foundation for the current performance reforms. I extend special thanks to my deputy Managing Directors, Directors, all other members of management and staff for the collective leadership that has helped to sustain high performance of NWSC since 2013. Without your acceptance and support of new, drastic and, sometimes, groundbreaking institutional reforms, there would not be a case study to write about. Specifically, I give special tribute to area managers and branch managers that spared some time to give exhibits of their experiences, which have helped to validate some segments of narratives in the case study.

I want to thank my family: my dear wife Mrs Annet Katussime Mugisha, my children (Silver G. Bakwatanisa, Edina Ainomugisha, Elina Bainomugisha, Enock Bamugisha and Annuciate Abomugisha and others whom we have adopted and lived with for a long time) for providing homely and emotional support. This provided great requisite emotional incentive when I was writing this book. I am, eternally, grateful to my parents, Mr Frederic Ngambagye and Mrs Paskazia Ngambagye (RIP), for a resilient and dedicated upbringing during the very difficult economic times of Uganda in 1970s and 1980s. To my brothers and sisters: I thank you, heartily, for special regards and prayers you have always extended to me.

Lastly, I want to thank IWA Publishing team for accepting to publish this book. Your publishing house has enormously contributed to my career growth in publishing peer-reviewed articles, books and book chapters. I want to thank Ms Mercy Timbitwire for providing editorial and proofreading support during the process of writing and compiling this volume. Special thanks also goes to anonymous reviewers for providing useful comments and inputs that enhanced the quality of the book. I want to thank Mahmood Lutaaya for his advice and input whenever called upon. Your invaluable work on the book outside office hours will forever be appreciated.

Chapter 1
Introduction

The water sector faces a decisive moment in countries throughout the world. More than 750 million people survive without safe water sources (worse still, over 2 billion people are experiencing intermittent water supplies), and over 2.4 billion live without safe sanitation. The next 10–20 years will be a period requiring a host of decisive actions for water management. It will be a period with serious challenges and risks, but also with significant opportunities, mostly in the provision of safe water and decent sanitation. The demand to allocate water efficiently to cities, industries, energy, food and ecosystems presents an unprecedented challenge to those developing and implementing policy. In tandem, it also presents a unique opportunity to create sustainable water management for the future. With increasing pressures from global change, coupled with existing unsustainable managerial practices, cities around the world will experience difficulties in efficiently managing less reliable water resources. In order to meet these challenges and to provide water security, there is the need for change and innovation in the water sector.

Delivering solutions for the complex water management problems that we face today and in the future requires a focus on bridging the gaps between sectors, and raising advocacy and awareness at a political level. It also

Author: Silver Mugisha
doi: 10.2166/9781789060256_0001

requires leading-edge technologies, targeted scientific research, new mindsets, and the right human competences, all to be combined with the best management practices. As a sector, there is need to embrace disruptive technologies and thinking, and adopt them at a much faster rate if we are to deliver the optimal benefit they can bring. We must 'take the bull by its horns' and measure up to the enormous challenges ahead.

1.1 THE WATER, ENVIRONMENT AND SANITATION SECTOR IN UGANDA

The water and environment sector of Uganda consists of two subsectors: the water and sanitation (WSS) subsector and the environment and natural resources (ENR) subsector. The WSS subsector comprises water resources management; rural water supply and sanitation; urban water supply and sanitation; and water for production. The ENR subsector comprises environmental management; management of forests and trees; management of wetlands and aquatic resources; and weather and climate.

The institutional framework for the sector consists of: the Ministry of Water and Environment with the Directorates for Water Development (DWD), Water Resources Management (DWRM) and Environmental Affairs (DEA); local governmental bodies (Districts and Town Councils), which are legally in charge of service delivery under the Decentralisation Act; and a number of de-concentrated support structures related to MWE, which are at different stages of institutional establishment, including Technical Support Units (TSUs), Water Supply Development Facilities (WSDFs), Water Management Zones (WMZs), and Umbrella Organizations.

The sector also has four semi-autonomous agencies: (i) the National Water & Sewerage Corporation (NWSC) for urban water supply and sewerage; (ii) the National Environment Management Authority (NEMA) for environmental management; (iii) the National Forestry Authority (NFA) for forestry management in the government's Central Forest Reserves; and (iv) the Uganda National Meteorological Authority (UNMA) for weather and climate services. There are also NGOs/CBOs (coordinated through UWASNET and ENR-CSO Network) and Water User Committees/Associations; plus the private sector (water and sanitation infrastructure operators, contractors, consultants and suppliers of goods). The activities undertaken in Sanitation and Water for Production (mainly focusing on agricultural and animal production) require close coordination with other line ministries including the Ministry of Health, Ministry of Education & Sports and the Ministry of Agriculture, Animal Industry & Fisheries.

The evolution of the water and environment sector in Uganda has been coupled with a number of reforms. There has been significant policy and institutional reform, bringing elements of performance management, performance monitoring and evaluation, multi-stakeholder coordination and decentralised management approaches. Understandably, there has been more emphasis on water supply than on sanitation coverage. On the water supply side, the concept of 'some water for all rather than all for some' has helped to navigate scarce resources. Sanitation coverage faces a number of barriers. These include: heavy resources required for infrastructure, obstacles in convincing people to connect to conventional sewerage systems, inadequate business orientation towards sanitation, poor physical planning in informal settlements, and inadequate incentives for public utilities to focus on sanitation.

1.2 PUBLIC VERSUS PRIVATE MANAGEMENT IN THE WATER SECTOR

In the 1990s, there was a significant debate in developing countries on what had to be done to rid public water utility companies of a plethora of problems they were facing. The problems included: poor customer care, rampant corruption and rent seeking, delayed maintenance of facilities, inadequate financing to cover operation and maintenance, political interference, and wasteful expenditure coupled with inadequate tariffs. These problems resulted in extremely poor service delivery characterized by unreliability of supply (often less than two hours per day), poor water quality and low service coverage due to limited investment financing. It was rare to find any water utility with a coherent strategic plan. Consequently, there were no performance measures on the basis of which accountability could be enforced. As a result, staff and managers enjoyed a lot of free-riding, with little regard to delivering value for the organisations in which they worked. There were no performance incentive plans in place, and employee promotion, let alone recruitment in the first instance, was largely based on patronage and technical know-who (rather than know-how).

Clearly, as can be seen above, the problem in most of the utilities can be summarized as lack of adequate finances for operation and maintenance (including depreciation) and capital investments, low operating efficiencies, poor customer care and demotivated employees. The debate was, therefore, on how to tap into the private sector to help address most of the problems above. The assumption was that the private sector (in contrast to the public sector) had an inherent incentive to bring investment capital and optimize operating efficiencies to maximize return on capital employed. Of course, the proposition was that the private sector's usual profit maximization mission would lead to significant efficiency gains, which would in turn yield improved service delivery to citizens. One of the concerns was the level of profits to be made, which was to be managed through contracts and other regulatory options. On the other hand, efficiency gains would be managed by incorporating performance measures/targets in contracts and subsequent regular monitoring and evaluation. The other concern was the risk of the private sector not delivering what it promised and whether the oversight bodies had the capacity to discern performance gaps and take remedial measures.

But what actually happened? The private sector solution came with a host of expectations and was, indeed, experimented with in several countries in Africa. These included: Uganda, Tanzania, South Africa, Mozambique, Cameroon, Kenya, Ivory Coast, Burkina Faso, Senegal, and Rwanda. The benefits in terms of leveraging private capital and improving efficiency gains have varied from country to country. Some countries still continue with the private sector management option, while others have abandoned it and returned to public sector management, by undertaking a host of attendant performance and governance reform initiatives. Uganda is one such example, where management contracts were piloted in Kampala City under the auspices of the NWSC in 1997–2004. The first was with a German firm, JBG Gauff Engineers, and the second was with a French firm, Ondeo Services Uganda Ltd[1].

The private sector in Uganda's water sector had several attributes. First, it brought a strong performance culture among indigenous senior managers and staff. Second, because the private sector operated hand in hand with the NWSC's public sector management in other towns, there was a strong sense of comparative competition, which helped to improve performance. Moreover, there was capacity building of NWSC managers in contract negotiation skills, performance monitoring and evaluation, and performance management.

---

[1]Mugisha S. and Brown A. (2010). Patience and action pays: a comparative analysis of WSS reforms in three East African cities. *Water Policy*, **12**, 654–674.

However, the private operator did not achieve contractual targets as expected. This, in turn, affected the value for money that should have accrued from efficiency gains. The private operator in Uganda faced a number of performance challenges. These included inadequate knowledge of the local operating environment, which affected the design of tailor-made solutions. For example, one of the targets in the contract was non-revenue water (NRW), which was mainly driven by illegal connections. The operator lacked tacit knowledge about the underlying casual factors: how it was created and how the problem could be solved. The local managers and staff, some of whom were complicit in the illegal connections, found no incentives to provide useful information to the operator's expatriate staff. Other challenges related to limitations in networking with key stakeholders and uncooperative native employees due to inherent sabotage behaviour.

Consequently, there were no noticeable performance advantages to private sector management over public management. That is why when the contract of the private operator ended in 2004 it was not renewed, and was replaced by public management augmented by significant new public management (NPM) tenets. The next section looks, in brief, at the practical aspects of NPM as applied to NWSC.

1.3 THE NEW PUBLIC MANAGEMENT CONCEPT

***Definition and Historic Perspective*:** New Public Management (NPM) is an approach to operating public service enterprises that is used in government and public service entities and agencies at both regional and national levels. The NPM approach aims to make public service more business-like and to enhance its efficiency by application of private sector management models. In other words, NPM focuses on the centrality of citizens who are customers of public services, in the same way that private sector management makes customer service central. In the 1980s, NPM reformers looked at a number of performance drivers to improve public service managerial efficiency. These included: use of staff incentives, decentralization approaches, enhancing commercial orientation, inculcation of market structure, and an emphasis on incorporating attributes of organizational behaviour in management practice. In water utilities specifically, emphasis was placed on performance management, customer care, delineation of an array of objective performance indicators, cost optimization, revenue enhancement and capital mobilization for infrastructure growth and maintenance. The ultimate objective hinged on improving efficiencies and more financing for service expansion. However, the reform processes in different countries faced different challenges. Most of these related to inertia to change from overly bureaucratic ways of working, corruption tendencies, compliance rather than business-like work cultures, political patronage and attendant cheap popularity, and misaligned organizational behaviour, among others.

***Broad Approach and Results*:** Consequently, the results of NPM reforms were, by and large, sub-optimal. Revenue collection efficiencies continued to be low, with government agencies not paying, illicit activities perpetuated by company employees in the field, water losses and poor customer care continuing to characterize service delivery. In some countries, utility managers, under pressure from development partners, introduced managerial reforms that they were not ready to cope with and were not, on their own, conversant with. With the help of 'flying' consultants, concepts like the Balanced Score Card (BSC) were introduced in performance planning and appraisal. However, because these concepts were not well conceived, the buy-in remained mundane and effective implementation remained far from the frontier. Other bodies were 'forced' into forms of governance arrangements like independent regulation, regionalization and restructuring without proper problem analysis. As a result, reform solutions were sometimes not well aligned with situational problems, thereby creating institutional redundancies, overlaps and huge cost allocative inefficiencies. In such cases, the main compelling driver for utility policy makers and managers was related to a desire to shed legitimate responsibility and put it into somebody else's hands without looking at the opportunity cost. The inverse approach, is of course, a partnering approach where someone else is seen as a helper who comes to give you support to deliver workload to fruition. This approach enables you to lead the way, as a change owner, and take responsibility thereto.

***Application Highlights to NWSC*:** Since early 2000 the approach to NPM reforms in the NWSC has been characterized by a strong customer and commercial focus, performance management, charismatic leadership, and quantitative staff incentive plans. As a result, the corporation has progressively witnessed employee alignment to a strong performance culture enabled by successive internally delegated contract frameworks. During the period 2010–2017 there was increased orientation to individual performance agreements (IPAs) to further strengthen individual staff productivity. In addition, during the latter period the earlier gains were consolidated through more focused strategic plans, designed around purposeful strategic priority areas (SPAs) with a longer planning time horizon of five years, rather than the traditional three-year corporate plans. With careful definition of SPAs, using a participatory approach and distilling guiding tenets from wider national policy documents (such as Uganda's National Development Plans, Vision 2040 and Manifesto aspirations), the strategic plans have become signposts of commitment to service delivery. Indeed, the '*water for all*' catchphrase has been holistically mainstreamed and become the official work song and attitudinal change driver. Strong orientations toward customer care and community engagement have also been mainstreamed, so that NWSC operates in a private-sector-like manner more than ever before. The recent characterization is that NWSC is a publicly owned institution operating like (or even better than) the private sector. This has been clearly demonstrated by the corporation being a two-time winner of '*best customer care*' provider by social media voters during the period 2014–2017, where the competitors included Uganda's best privately operated companies. The centrality of customers and citizens has also been tested through periodic 'customer parliaments' (structured and facilitated gatherings) where customers voice their satisfaction and concerns in an open and transparent manner.

***Observations and Lessons*:** Using the snapshot overview of NPM above, one can make a number of observations and draw lessons. First, the NPM approach presents great prospects for efficiency improvements in public service delivery and hence increased capacity to create cost savings and recipes for capital investment finance mobilization. Equally, however, there are many pitfalls and the potential for policy makers and utility managers to *pretend* to be delivering services under the guise of NPM reforms. They will do this mostly to divert shareholder (political) interest to create value for citizens. The challenges of public enterprises, especially in developing countries where efficiency delays are significant, remain a daunting task to deal with. There is much that those charged with service delivery can do through strong leadership, enabling change and setting the scene for a new paradigm for service delivery. A lot has to do with '*holding the bull by its horns*', applying a do-it-yourself approach to create tailor-made local solutions. You cannot privatise your problems, especially where you have superior informational advantages to address issues. It is now more obvious that each country (or region) needs to design its own context for NPM reforms that directly address prevailing performance problems. There are no universal solutions for a host of locally diverse problems. What can be sought, however, is the capacity to deal with change, seek innovativeness and deep appreciation of technological advances and new global opportunities and threats.

1.4 EVOLUTION OF NWSC PUBLIC MANAGEMENT

Some highlights of NWSC public management have been briefly presented in the preceding sections. This section gives a more coherent chronology of the evolution since the inception of NWSC in 1972.

***Formative Period (1972–1986)*:** As presented by Muhairwe[2], this period was characterized by the creation of NWSC in 1972. However, there was not much progress in terms of improved service delivery either horizontally or vertically. By 1986, NWSC was only operating in five towns of Uganda and service delivery was heavily constrained by political interference, rent seeking by employees, and inadequate capacities of managers and staff who were responsible for infrastructure management. Moreover, there was not enough resources (human and financial) to maintain and expand the system for effective service delivery. There were, however, steps that (though slow) were progressively taken to improve things. For example, some graduate engineers were recruited to enhance technical managerial know-how. Unfortunately, the speed of improvements fell far below the rate of growth in demand for water and sewerage services. By 1986, most homesteads in towns where NWSC was operating had taps but little or no flowing water. Service reliability was less than three hours per day. Service coverage (including those connected on empty pipes) was under 15% of the target population. Water losses were estimated at over 80% of water produced, and revenue collection efficiencies were below 40% of the few customers that were being billed. Maintenance of facilities had been delayed for some time due to inadequate financial resources and many were in need of significant rehabilitation and remediation. There was not much of what one would call customer and commercial orientation, both from the point of view of staff work attitude and the technology options available to offer customer services. In a nutshell, 'service delivery' during this period was not something worthy of the name by any definition.

---

[2]Muhairwe W. T. (2009). Making Public Enterprises Work. IWA Publishing, London, UK.

***Revitalising the Systems (1986–1998)*:** The period between 1986 and 1998 marked a step change in the performance of NWSC. The number of operational towns under the corporation increased from five to eleven. More pipe infrastructure, water treatment systems and office premises were constructed, mostly with support from donor funds. At the same time, priority infrastructure rehabilitation was undertaken to restore production and pipe distribution system carrying capacities. Management systems were introduced, namely: computerized billing system, block-mapping/GIS, computerized financial management system, water quality management systems, procurement and logistics. In addition, a corporate planning culture was introduced while the legislative arrangements establishing the corporation were reviewed and strengthened. The tariff was also reviewed to better reflect the cost of running business, albeit not being cost-recovery. Furthermore, professional staff were recruited and a training centre was constructed to support continuous learning through refresher and short courses. Staff were deliberately sent for further studies outside the country as a way of building sustainable and modern managerial capacity. As a result, service coverage increased from about 15% as at 1986 to about 48% in 1998. Pipe network length increased from about 500 km to about 2000 km. Annual revenues increased from about US$ 2 million to about US$ 20 million, albeit at low collection efficiencies of about 60%. Hours of supply increased to more than 15 hours per day. Water losses reduced from over 80% to about 60% of water produced, though this was still very high. Staff productivity increased from over 60 staff per 1000 connections to about 36 staff per 1000 connections, as the total number of connections increased from under 10,000 in 1986 to about 50,000 in 1998. Clearly, there was remarkable business growth, although customer and commercial orientations remained mundane. As a result, return on investments undertaken remained very low, considering that the corporation was still incurring deficits of about US$ 4 million before depreciation. The NPM reforms undertaken had not yet been fully effective.

***Harnessing the Systems (1998–2011)*:** This period marked a change in top leadership, including the Managing Director/Chief Executive Officer and the Board of Directors. As a result, there was marked improvement in commercial, customer orientation and organisational behaviour. Under this era, new performance frameworks were introduced, ranging from simple memoranda of understanding between top management and staff teams to detailed internally delegated performance contracts. These frameworks stipulated obligations, performance targets, incentive plans, action plans and monitoring and evaluation arrangements. Vigorous cost containment measures were put in place setting the scene for cost consciousness among managers and staff. In addition, customer care and integrity reassurance systems were introduced. Project management was more transparent, with marked reduction in project time and cost overruns. More infrastructure (treatment plants, pipe networks and office premises) were constructed. At the same time, priority rehabilitation works were carried out, while offices were refurbished and improved in ambience. It is at this time that the concept of transparent and open offices was introduced. The tariff system was improved further through introduction of an indexation mechanism that has hitherto protected the tariff against erosion by exogenous factors like inflationary effects, foreign exchange and power costs.

During the same period, there was increased momentum for staff to go for further studies including undertaking PhD programmes in various disciplines. As a result, there was significant performance turnaround. Service coverage increased from about 48% in 1998 to 75% in 2011. The number of operational towns increased from 11 to 23. Pipe network length increased from about 2000 km to 6500 km while customer connections increased from 50,000 to 272,000. Water losses reduced from about 60% to about 33%. Annual revenues increased from about US$ 20 million to about US$ 52 million while profits before depreciation increased from a deficit of US$ 4 million to a surplus of US$ 12 million. Moreover, the collection efficiency increased from about 60% to over 95%. The supply reliability increased from about 15 hours per day to over 18 hours per day. The staff productivity increased from 36 staff/1000 connections to 6 staff/1000 connections. Clearly, the performance results suggest that, during this period, the NPM reforms began to pay-off in terms of efficiency gains, customer service and return on investments undertaken.

***Consolidation and Focus on Service Delivery (2011–2017)*:** The period between 2011 and 2017, as depicted in Tables 1.1 and 1.2, and staff exhibits #1–4, has been characterised by consolidation and rapid expansion of NWSC's geographical footprint from 23 towns in 2011 to 218 towns in 2017. The current public policy is 'ensure water for all' through a combination of NWSC internally generated resources and targeted government investment subsidies. This has in turn increased the urban population served from 2.5 million to 6.3 million. The service coverage, however, did not grow at a high rate due to low coverage of new towns that has mathematically affected the average value. The pipe network

Table 1.1 Performance trends for 218 towns (1998–2017).

Performance Indicator	1998	2011	2017
Number of NWSC towns	12	23	218
Pipe network length (km)	2000	6500	12,500
Water service coverage (%)	48.4	75.1	78.2
Total connections	50,826	272,406	530,000
New connections per year	3317	25,633	40,712
Proportion of metered accounts (%)	65.3	99.8	99.8
Staff per 1000 connections	36	6	6
Collection efficiency (%)	60	98	93
Non-Revenue Water (%)	60%	33	31
Turnover (billion Ush/year)	21	132	321
Operating surplus/deficit (before. dep) (billion Ush)	8.0 (−)	30 (+)	71(+)

length has increased from about 6500 km to 12,500 km due to a combined effect of take-over of already built infrastructure and new infrastructure constructed using internally generated funds. The total connections have increased to 530,000 due to high new connection rate of about 40,000 per year. On the financial side, annual revenues have also almost doubled due to increased business growth. With strict cost containment measures, the

Table 1.2 Performance trends for 23 towns (1998–2017).

Performance Indicator	1998	2011	2017
Number of NWSC towns	12	23	24
Pipe network length (km)	2000	6500	8340
Water service coverage (%)	48.4	75.1	85.3
Total connections	50,826	272,406	461,020
New connections per year	3317	25,633	37,095
Proportion of metered accounts (%)	65.3	99.8	99.9
Staff per 1000 connections	36.0	6.0	4.9
Collection efficiency (%)	60	98	93
Non-Revenue Water (%)	60	33	30
Turnover (billion Ush/year)	21	132	311
	8.0 (−)	30 (+)	89 (+)

profits before depreciation have also significantly increased, creating enhanced internal capacity to carry out infrastructure expansion and timely maintenance and rehabilitation. Table 1.2 specifically shows that the take-over of new town dampened some performance indicators. The network length in the original 23 towns has been growing at an average of 400 km per annum over the four years (2013–17) compared to the previous rate of 80 km per annum. Non-revenue water in the original 23towns has also continued to go down, meaning that the take-over of other towns did not affect maintenance and remediation efforts. As already pointed out, overall, commercial performance has improved suggesting that the technical capabilities of operators in new towns has improved through cross-pollination of staff skill sets.

EXHIBIT #1: FROM CHARLES OKOUNZI (2018), NWSC GENERAL MANAGER

My summary of the key activities and lessons with regard to what has happened since 2013 when we got a new leadership includes a number of developments, namely:

Implementation of Infrastructure Service Delivery Plans (ISDPs) and 100% Service Coverage Acceleration Project (SCAP100): The launch of the ISDPs, to me, was a game changer. By then, I had just entered into area management and we extended water coverage at a rapid pace. With substantial team effort, we laid over 90 km in one year (yet doing 3 km was previously considered to be a very serious achievement). Many parts of Tororo and Busia benefited and the resultant impact was that more communities and institutions got connected to the NWSC grid and the revenue base almost doubled. We are doing the same with SCAP100 and service coverage is increasing like nothing. More villages are getting connected and enjoying our services at a momentum never comprehended before.

Delight of performance-based reward: In this regard, prioritizing staff welfare has been a core driver for the forward change in performance. In the last two years my entire team and I have received salary raises three times due to good team performance. This was something unimaginable in the past. I can say my team, if not the best, is one of the best motivated, delighted and energized in NWSC. Besides, we have received other financial rewards. Indeed, as we are growing the business and stepping up performance, we are also growing our pockets.

Training and engagement: I will never forget the day I flew to Dubai to attend a strategic leadership training session with the MD. Apart from every nice and memorable experience, I felt much loved and abundantly appreciated as an employee like never before. This made me double my efforts thus doubling the performance of my team. In fact, I committed to never fall short of achieving performance targets.

Customer and stakeholder connectivity: The introduction of the Information, Education & Communication (IEC) strategy by creating awareness and disseminating vital service related information turned NWSC into a caring and well-meaning organisation, rooted in the community. This helped to build-up supportive behaviour among the community and stakeholders in water service delivery. With the public trust and loyalty, the pendulum swinging in favour of NWSC, an upsurge in performance is assured.

Visionary-servant leadership: Having a leader with a servant's heart, with an extremely strong and uncommon desire to serve others is a truly invaluable asset, which has caused a step change in performance. Never before have I seen a leader put his needs on the back burner in order to move an organization forward. In the last four years, I have seen our MD make time to meet with all staff and stakeholders to ensure everyone has a true stake in the direction of the corporation. He has created a team-like atmosphere where everyone contributes to the vision and mission of the corporation. Our MD is willing to meet people wherever they are which builds trust and collective efforts of everyone within the corporation has produced positive results.

EXHIBIT #2: FROM JULIET NAKANJAKO (2018), NWSC WATER QUALITY OFFICER

This is my brief take on what has contributed to step change in the corporation in the last five years:

Joint problem solving: This has been enhanced by the creation of WhatsApp groups for each directorate and department. Here, all area activities are reported on a daily basis, most importantly, during the times of incidences or service delivery interruptions. The social groups are used as a communication channel for guidance about corrective actions.

Managing Director's involvement in the corporation activities at all management levels: The existence of the MD on all departmental social media platforms enhances the performance of departments because he involves himself in all these activities through; posting customer concerns, providing guidance on performance improvements and communicating executive management plans/priority areas.

Strengthening customer relations: The postings of customers' compliments and complaints via Twitter has strengthened customer relations because of the quick response time that is used to attend to customers and the feedback provided. In the long run, this has also enhanced the corporation's image.

Reliable laboratories: The regional and area laboratories have been equipped and are hence capable of conducting all the necessary water quality tests. The staff manning the laboratories are also equipped with the necessary knowledge and skills.

Alignment of staff job descriptions to the corporation strategic direction: Through the drafting of Directorate Action plans which also stem from the corporate plan, departmental action plans are also drafted providing room for drawing staffs individual action plans in a bid to execute the departmental key result areas. This has enabled staff to execute their activities in relation to the corporate strategic direction and perform tasks with a purpose upon which they are evaluated during appraisals.

Stakeholder's engagement: The level of stakeholder engagement has been enhanced most especially for the people's representatives (MPs, RDCs, local committees and council etc.). There is significant involvement and in the process, key corporation plans are communicated. In the end, there is an opportunity for customers who are not able to use the social media to give feedback to the corporation's managers.

Enabling the adherence to SDGs by the government: The corporation has successfully complied with Government's ambition of adhering to goal 6 (and specifically target 6.1) of the SDGs, which aims at achieving universal and equitable access to safe and affordable drinking water for all by 2030. This contribution has been enhanced through initiating investment programmes like the SCAP100 that aims at increasing service coverage to 100% by 2020.

Capacity building: This has been enhanced through creation and equipping of the new mechanical training centre located at Ggaba waterworks where most staff are trained. This has tended to increase their knowledge, skills and abilities (KSAs) to execute their duties. The number of staff undertaking benchmarking in foreign countries has also increased. Employee motivation has also been enhanced through salary scale revisions and periodic promotions.

EXHIBIT #3: FROM ANDREW MUHWEZI (2018), NWSC WATER PRODUCTION MANAGER

I would characterise what has happened to cause a stepped change in performance in the last four years in a number of ways. First, bold decision making by NWSC leadership to improve service quality (especially reduction of 'dry' zones) through major infrastructure upgrades and network extensions. The advent of organizational restructuring is also considered a motivator in the sense that it opened up new positions and encouraged staff to work harder so as to improve their prospects of getting to those positions. Management embracing modern real-time communication platforms like WhatsApp has enhanced sharing performance information on a day-to-day basis across the organization and this has facilitated internal benchmarking. In addition, bold initiatives by NWSC top leadership to challenge the status quo in all performance indicators with a focus on improving satisfaction of both internal (staff) and external customers. Also, inexhaustible quest by NWSC leadership to promote and encourage innovative thinking and approaches in all NWSC operations.

EXHIBIT #4: FROM ADAM NABUSAGU (2018), NWSC BRANCH MANAGER

In my view, the five key activities that have stepped up performance of NWSC in the last four years have been varied. The rolling out of the

SCAP100 program which is aimed at extending safe clean water to over 12,000 villages across the country. The rolling out of the water stabilisation plans which has seen most of dry zones eliminated. The increase in the geographical coverage from 24 towns in 2013 to over 225 towns, giving NWSC a national presence. Increase in infrastructure growth from an annual average of 80 km five years ago to over 1000 km of water pipe network extended per year across the country. Finally, a delighted water force which has led to an improved customer service.

1.5 SCOPE OF THE BOOK

This book gives a detailed account of approaches and strategies that were applied by Board and management of NWSC after 2011 to ensure effective business continuity and expansion. It is a useful primer building on the detailed account given by Dr Muhairwe[3] in 2009. The book presents a strong case study, given that the leadership prior to 2011 (with Dr Muhairwe as Managing Director for the period 1998–2011) was largely regarded as charismatic and transformative. Many NPM reforms were introduced as briefly summarised above and significant results were achieved. It is interesting for policy makers, utility managers and academicians to know what happened after this era. In writing this book, we use a combination of quantitative and qualitative methods to present performance trends and corresponding explanatory underpinning narratives. In order to give context and a holistic perspective of approaches and inputs used by NWSC, we use a balanced scorecard framework. Consequently Chapter two looks at the internal perspective detailing some central technical inputs and tools that have been instituted and applied to operational excellence with innovation as well as being socially responsive. Chapter three delineates the detailed underpinning people-oriented perspectives and processes that have been applied to support NPM reforms. Chapter four, on the other hand, deals with detailed tools and inputs from a financial perspective. Chapter five details inputs under the customer and community perspective. Lastly, Chapter six presents summary observations, lessons and conclusions. Where

[3]Muhairwe, W. T., *ibid.*

appropriate, the chapters are interspersed with visuals, pictorials, specific performance trends and proverbial narratives.

It should be noted that after the contract of Dr Muhairwe as the Managing Director of NWSC ended in November 2011, there was a transition led by Eng Alex Gisagara up to August 2013, when a new Managing Director, Dr Eng. Silver Mugisha, was appointed by the Board of Directors. The search for a replacement of Dr Muhairwe was quite a prolonged process, marred with some challenges, although this is not the main subject of this book.

The appointment of Dr Mugisha as the Managing Director came against a backdrop of much expectation from Government. Generally, there was feeling from the political leadership that much as the corporation had registered significant efficiency gains in the last 15 years, there was not much work done in expanding the network and geographical coverage. In addition, there was inadequate attention and collaboration on environment management issues and yet this aspect was hugely affecting the existence of raw water sources. The third area of concern was inadequate collaboration with the political establishment – with a need to balance technical work and political expectations. In addition, internal staff engagement, participation and connectivity needed improvement to create a better positive work environment and sense of belonging. These four areas: national service coverage; more emphasis on environmental management; increased collaboration with political management and increased staff engagement formed the basis for redefining the strategic direction of NWSC under the new leadership.

1.6 TARGET AUDIENCE

This book volume is intended for policy makers, utility managers, academicians and researchers. The case study herein presents useful knowledge that those intending to design effective strategic directions to impact sector utility performance can obtain and apply. The knowledge is practical, tested and fit-for-purpose, mostly in developing country contexts. Institutional and organisational development consultants will also find it useful in designing practical solutions for utility problems. The use of a balanced score card (BSC) framework presents a good basis for academicians and researchers to discern a wide array of propositions for scholarly research in issues of utility management. Most importantly, the constituent tools and inputs under each perspective warrant further research so that we can have tested evaluative frameworks specific to developing country contexts. The tactical approaches underpinning performance improvement can also act as useful catalysts for other utility practitioners in similar operating environments as either blue print solutions or triggers of innovative approaches. The inherent performance risk triggers and remediation measures taken are also useful areas of interest for utility practitioners.

Chapter 2

Internal perspective

This chapter looks at measures that were taken to ensure operational excellence as well as being socially responsive. In other words, we present inputs, tools and strategies that relate to four thematic areas, namely: operational excellence, process innovation, strategic alliances and socially responsive initiatives. Staff exhibits #5 and 6 show some elements of the internal perspective on these activities in the period from 2013 to 2018.

EXHIBIT #5: FROM ANDREW SEKAYIZZI (2018), NWSC GENERAL MANAGER

Over the last four years, I can characterise what has been going on in the corporation to improve technical performance in a number of ways. We have seen charismatic leadership which allows people to innovate, for example our in-house billing system and other applications. Taking calculated risks and refusing to be slaves of established norms and processes, focusing instead on output and results for our people. For example, use of direct labour to extend services to people more cheaply. We have ensured that we stay connected to and responsive

Author: Silver Mugisha
doi: 10.2166/9781789060256_0019

to the communities and people we serve. We have had performance management and an incentive framework that rewards all for performance – staff, therefore, see NWSC success as their own success. We have had strong strategic thinking and building of goodwill with government and local leaders. Emphasis on development rather than consumption expenditure. Visionary and bold leadership, which has seen NWSC expand to over 225 towns against predictions of doom and collapse.

EXHIBIT #6: FROM FRANCIS BBALE (2018), NWSC SENIOR ENGINEER

Below are some of the activities and key lessons that I personally relate with the observed step change in NWSC in the last four years. We have had transparent leadership, with open-door management style, sharing realistic performance targets and performance results with all staff. This has instilled a sense of continuous individual performance to always meet the set targets. The strong strategic business planning orientation has also been paramount. In this regard, continuously, business streams have put in place a framework for informed decision making, conditional assessment of the utility, re-emphasising corporation's vision, core objectives, performance strategies and continuous engagement of staff for continuous performance improvement. We have also seen practical organizational approaches: management has been involved in exercises that actively engage staff for individual and Corporation growth. The most touching one is when the MD himself made a round to all units, sections, branches, and areas preaching a gospel of staff career growth (pursuing further studies, enrolling in short courses to improve their personal skills) and also continuously working hard to earn recognition. This has been a positive way of motivating staff and many stepped up their performance following this action. In addition, NWSC has invested in capacity building programmes for example: technical training programmes to staff at IREC and Gaba (practical training ground). Furthermore, staff have been taken out of the country for benchmarking and training. This has greatly improved staff competence and exposure hence improving their capacity in critical thinking as far as providing practical solutions to prevailing problems is concerned.

2.1 OPERATIONAL EXCELLENCE

During the period 2013–2018, under the leadership of Silver Mugisha as the Managing Director, NWSC put in place a number of strategies, technological inputs and tools that helped to improve operational excellence. This section gives a detailed account of some of the salient measures. Staff exhibits #7 and 8 show some activities contributing towards excellence.

EXHIBIT #7: FROM HELLEN EJANG (2018), NWSC BRANCH MANAGER

In my view, a number of groundbreaking activities have greatly contributed to the corporation's operational excellence in the last four years. These include: high level of investment in pipe extensions, construction of several public water points for urban poor communities, customer focus and stakeholder management. This has been enabled through good leadership of Dr Silver Mugisha that has encouraged high level of staff integrity, career growth and staff development. Furthermore, the restructuring exercise that led to introduction of regional management by highly qualified and experienced leaders has been key.

EXHIBIT #8: FROM ANNET KUKUNDA (2018), NWSC BRANCH MANAGER

Operational excellence of NWSC has had a number of contributory factors. These include: target oriented performance improvement programs which encourage competition among regions, branches and territories; able leadership that involves participation of staff in decision making at all levels; in-house IT systems and applications like CRM, a new billing system that saved the organisation a lot of money and improved the analysis of the KPIs; competent leadership with orientation towards staff development through training both in-house and out of the country and experience-sharing platforms for best practices; emphasis on customer delight and satisfaction by coming up with themes that are customer centred; strong emphasis on business growth.

2.1.1 Restructuring of management of operational areas

As highlighted in the previous chapter, in 2013 NWSC was operating in 23 towns in Uganda. The new Strategic Direction (2013–2018), approved by the Board and Minister of Water and Environment set an ambitious agenda to expand operations to over 80 towns by 2018. This was directed at reaching more Ugandans, through massive pipe extensions and water supply stabilisation to increase supply reliability. The main operational challenge associated with this strategic agenda was coping with operational excellence, making it even better. Thus, immediately after the approval of the Strategic Direction, management embarked on a restructuring exercise to create structural arrangements that would cope with rapid geographical expansion. Consequently, NWSC was divided into four operational regions: North-East, South-West, Central and Kampala City. Each of these regions was to be headed by a regional manager with a core team of secretariat staff. This was a marked improvement from the previous arrangement where all operational areas were managed under one operations manager. In tandem, other core operations were also regionalised, following the same restructuring formula. These included: static plant maintenance, water quality management, customer block-mapping/GIS, human resources, legal and land acquisition services. In order to create a corresponding monitoring and evaluation structure at headquarters, the organisational structure was also strengthened by creating positions of three Deputy Managing Directors, six Directors and a number of supporting senior managers. Exhibit #9 contains examples of some of the concerns of staff regarding restructuring.

EXHIBIT #9: FROM STEPHEN WALAKIRA (2018), NWSC AREA MANAGER

Staff can be worried about effects of restructuring. After implementation challenges were addressed, staff tend to raise impact concerns. For example: Is the change making a difference? Are we making progress? Are things getting better? Staff with impact concerns were interested in the relevance and payoffs associated with the changes. NWSC leadership focused on sensitising staff about benefits of restructuring. The good news is that leadership did a good job, this is the stage where staff tried to market themselves on the benefits of the change based on the relative merits of the performance results to be achieved. Leadership shared success and ensured that change is making a positive difference. The changes in the organisation structure brought positive impact results and clear measurable success is being witnessed.

The restructuring exercise had a number of advantages and some challenges. First, it significantly improved supervision and coordination of operations in areas. More functions like digital marketing, public relations, water quality management, static plant maintenance, legal services and land acquisition, applications development, office support services, corporate and customer relations management and risk management were given greater prominence. In addition, the structure created better avenues for succession planning and career progression. As a result, managers who were eventually internally selected to occupy newly defined positions were more elated and motivated. In the latter regard, positions were filled through a competitive process, allowing for career growth. These actions eventually resulted in improved staff productivity, commitment and work output. However, there were some challenges related to costs of restructuring which had to be managed through a phased approach to implementation of the structure. Furthermore, staff who did not make it through the competitive process were disappointed, but were encouraged to work harder, improve their competences and skills while targeting the unfilled positions and future restructuring possibilities. In other words, the situation had to be handled through careful change management that was characterised by counselling, fair dealing, transparency and objectivity.

Apart from regionalisation, management has created more commercial branches and sub-branches to bring services closer to the customers. This has been in addition to creating more convenient information technology (IT) solutions that will be discussed later. The IT solutions make customers conduct business seamlessly, but branches bring in a human touch and enable other technical operations that require a physical interaction. In these branches, management emphasises excellent customer service, ambience and, most importantly, advocacy for future expansion programmes. These operational outlets have increased collaboration between the communities and NWSC, which has in turn increased a sense of ownership and belonging.

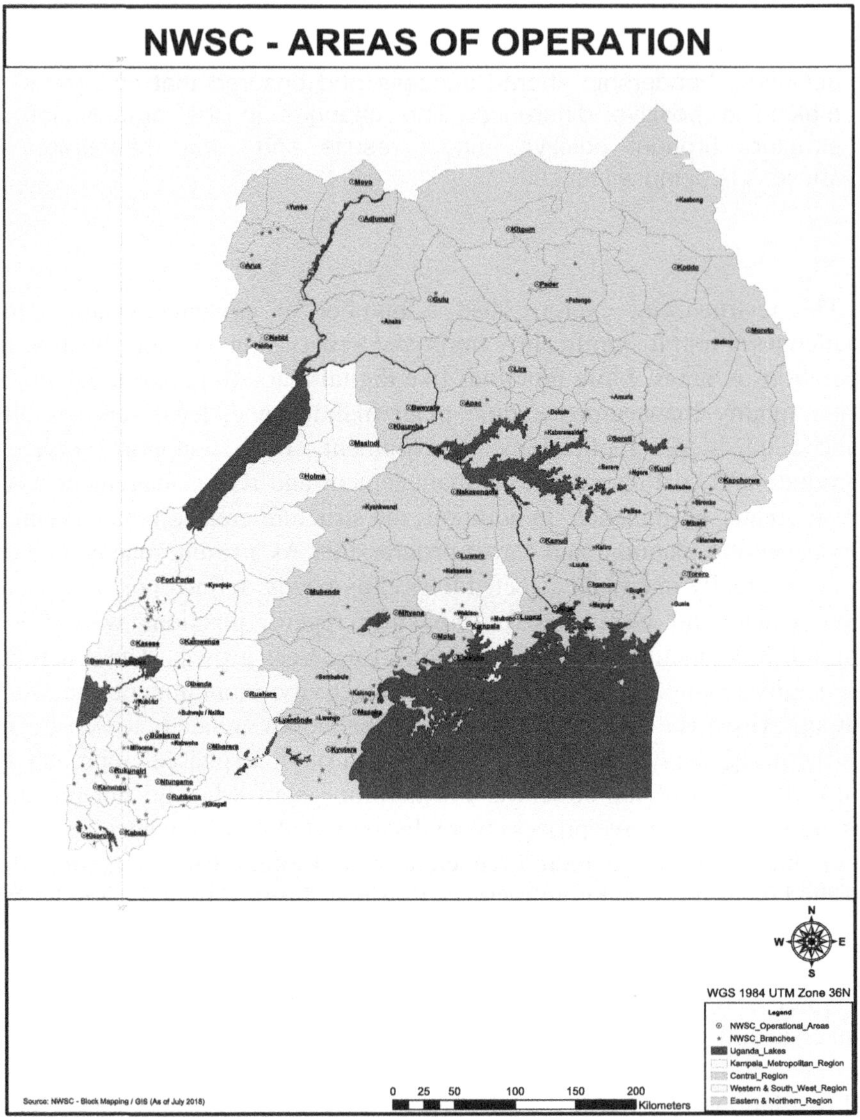

Figure 2.1 Map of NWSC regions and operational areas as of July 2018.

2.1.2 Performance contracting

It should be noted that NWSC has had a long history[1] of internally delegated performance contracting since 1998. During the period 2013–2017,

[1]Muhairwe, W.T., *ibid.*

however, the new management came up with a number of improvement measures. The concept of individual performance agreements (IPAs) was introduced, based on group action plans. Previously, there were what were termed 'one minute goals' (OMGs), which were not systematically linked to group action plans. As a result, productivity management was hard to enforce in a holistic manner. Moreover, OMGs did not have smart and sharp targets and hence their monitoring and evaluation was quite problematic. With the advent of more refined IPAs, it has been possible to carry out regular individual staff appraisals and give feedback in a more transparent and objective manner. During the same period, departments and directorates are required to prepare annual action plans, deriving from the Strategic Direction, detailing strategies, activities, targets and financial projections aimed at achieving annual corporate targets. The action plans are signed off as departmental and directorate commitments, on the basis of which periodic monitoring and evaluation is carried out by the performance monitoring department. The progress of implementation of annual action plans is also fed into the preparation of periodic performance reports to the Board of Directors for effective oversight and strategic guidance.

Another major ingredient of performance contracting that has improved operational excellence in terms of cost efficiency is a strong cost containment strategy at the negotiation stage of financial projections and management fees for areas. A lot of benchmarking is carried out prior to negotiation of fees and area managers are asked to keep management fees at thresholds that optimise cost efficiency. In this regard, headquarters-based performance analysts carry out comparative analyses in similar operating conditions and area managers are tasked to justify any cost escalations. While doing this, it is ensured that system maintenance and upkeep is not compromised, while consumptive expenditure that does not add value is minimised. Because of this approach, significant cost savings have been made and funds leveraged for capital investments. As a result, a strong cost-consciousness culture has emerged. That explains why the operating surplus has been on the rise since 2013 despite salary adjustments, huge expenditure on delayed maintenance, and the take-over of unviable operations in many towns that require heavy subsidies.

During the same period, the incentive plans have been improved by incorporating new parameters that reflect emerging performance areas that require improvement. For example, more weight is now given to improvement of business growth and optimisation of costs through

Figure 2.2 Members of NWSC executive management team signing their Individual Performance Agreements (IPAs), at the NWSC International Resource Center (IREC), Kampala, October 2014.

Figure 2.3 NWSC executive management team exchange the IPA document with the managing director Dr Eng Silver Mugisha (right) as a commitment to deliver the agreed targets, NWSC-IREC, Kampala, October 2014.

increased cash operating margins. The incentive plans are also more objective and reward performance whenever there is incremental effort. This is a major deviation from previous arrangements where maintaining high base performance was not rewarded. Exhibit #10 shows an example of staff perceptions about the reward system, which had tended to demoralise operational teams to the detriment of service delivery. In addition, there are periodic performance evaluations which are carried out on regional basis, thereby reducing operational costs while keeping performance momentum among staff. In such regional evaluation workshops, customers and community leaders are invited and asked to air out their concerns and feedback through facilitated barazas (public meetings that serve as community level accountability forums).

EXHIBIT #10: FROM SAM APEDEL (2018), NWSC PR MANAGER

There have been tremendous improvements through performance management and reward systems: through an incentive framework that rewards all. The workers see NWSC as *their* organisation and see its success as *their* success.

2.1.3 Water loss reduction

Water loss is one of the critical areas that affects the operational excellence of any water business enterprise. It is normally visualised in two forms: physical losses and commercial losses. Before 2013, NWSC did not have a rigorous approach to water loss reduction, although it had been targeted as low-hanging fruit in the initial reform – where there had been some success (as Table 1.1 shows reduction from 60% to 33% between 1998 and 2011). Nevertheless, Non-Revenue Water (NRW) was as high as 40% in Kampala (e.g.,) and there was less predictability regarding remediation measures. Consequently, the Strategic Direction (2013–2018) envisioned NRW reduction as one of the priorities.

It had always been clear that a lot of water was being lost through illegal consumption and tackling this problem would require deliberate effort. As such, the new management, working with security agencies (police and internal security organisations) came up with a series of water loss prevention programmes in all areas. The programmes involved extensive

Figure 2.4 NWSC management team together with the officers from the Uganda Police at the Launch of the Water Loss Prevention Unit (WALOPU) a partnership between NWSC and the Uganda Police to fight Illegal Water Use, at the Central Police Station (CPS) Kampala, August 2014.

Figure 2.5 A culprit caught and handcuffed by the WALOPU team and transferred to the Police post set up at the NWSC offices at NWSC offices on Plot 43/49, 6th Street Industrial Area Kampala. November 2014.

use of covert approaches to detect users (sometimes using incentives paid to informers) and overt approaches to arrest culprits. The culprits are fined, in accordance with the laws of Uganda, and asked to pay, either in whole or in part, through arrears performance agreement (APA) frameworks. In this connection, management has worked with the Judiciary of Uganda to establish a separate Utilities Court to handle emerging criminal cases in a speedy and fair manner. Of course, for a utility that is customer centred, taking customers to courts of law is not pleasant. Management, therefore, uses a lot of plea bargaining to resolve conflicts instead of escalating them. At the same time, publicity is used to 'name and shame' culprits to deter the proliferation of such illegal acts.

Another form of water loss in NWSC has been through the proliferation of pipe cuts caused during roadworks. In such cases, it matters a lot how utility staff react to prevent water wastage and, at the same time, ensure supply continuity for customers. The approach of the new management has been that there should be close collaboration with roadworks contractors to avert pipe cuts and water loss, and ensure that if they happen they are controlled in a timely manner. Given that the level of preparedness of roadworks contractors to repair pipe cuts is normally low compared to NWSC, the strategy is always to let the utility staff repair and charge the contractor, based on negotiated market rates of materials and labour used. This has helped to rationalise this activity and ensure supply continuity with minimum interruptions. Utility staff are strictly monitored to ensure compliance, which has helped to build resilience levels in this regard. Apart from causes due to roadworks, pipe cuts also arise from damage to exposed pipes due to soil erosion and/or laying pipes at shallow depths at the construction stage. During 2013–17, a tactical strategy called the *S*ervice Pipes *Wa*ter *Lo*ss *P*revention (SWALOP) programme was launched to re-lay all exposed pipes at depths of at least three feet to prevent surface damage by external factors. SWALOP has had significant positive impact on reduction of physical water losses.

A further source of water losses has been through meter inaccuracies due to old age, physical tampering and inappropriate technology. Other losses are due to mistakes (whether by omission or commission) by meter readers in the field. It was commonplace for meter readers to sit under the tree and forge meter readings; others would deliberately under-read the meters, while seeking rents from customers. In order to deal with these issues, management intensified meter testing and field audit activities where exceptional consumption patterns were observed. In addition, meter readers and their immediate field managers have routinely been given warnings that meter-reading irregularities would attract tough sanctions, including employment terminations. Amongst administrative measures taken, meter readers have been transferred more often to avoid getting used to customers in the same place and establishing 'fishing' grounds. A lot of sensitization work, including highlighting anti-corruption/bribery messages, was carried out. In this regard, telephone contact details for reporting bribery seekers were put on the body of all field vehicles. The extensive use of customer surveys from the Managing Director's office, specifically inquiring about field staff who solicit for rents/bribes while carrying out field activities also helped to send warning signals. In these surveys, the quality of service offer is also assessed and feedback given to concerned managers for immediate action.

2.1.4 Water supply stabilisation

Water supply stabilisation is another area that needed attention to improve operational excellence. As of 2013 most supply areas, especially in Kampala City, had acute intermittent supply resulting from undersized pipes, low pressures, blocked pipes, stuck service valves and malfunctioning or inadequate air valves. In addition, unreliable grid power

Figure 2.6 Pipe water network reinforcement in Kampala along Ggaba-Namasuba transmission main aimed at improving water supply reliability in Kampala, March 2016.

was another problem in most NWSC operational areas. A programme designated the Water Supply Stabilisation Programme (WSSP) was designed in line with the strategic direction to address this operational challenge. The necessity of the WSSP was evident from the rampant outcry from customers about supply unreliability, which was adversely affecting their loyalty and willingness to pay. As such the brand image of the corporation, as suppliers of 'dry zones', was increasingly being impacted.

The WSSP was a major corporate programme for all areas and was therefore designed and given sufficient financial and human resources in line with the Strategic Direction. The main activities of the programme included pipe network reinforcement by laying bigger parallel mains, rehabilitation and expansion of water production systems, installation of standby power generators, and installation of miscellaneous network appurtenances (service valves, washouts, air valves, fire hydrants, among others). The WSSP has been one of the most successful programmes, and has increased supply reliability, especially in Kampala City from an average of 15 hours of water supply per day to over 20 hours of supply per day, with most areas getting 24 hour water supply 7 days each week. Areas that used to be 'dry zones' now enjoy reliable water supply. Where installing standby generators was not technically viable, as in Kampala City, there was increased collaboration with the power company to establish a dedicated line of power supply.

Figure 2.7 Contractor's staff connecting the DN 700 Ggaba-pipe water transmission main to boost water supply reliability in Kampala. November 2015.

As a result of water supply unreliability, which was becoming worse with increasing new connection hook-ups and hence increased demand, the customer satisfaction index (percentage of satisfied customers from a random sample) had deteriorated to about 79% in 2013. As of 2017, the customer satisfaction index has steadily increased to 86%, partly as a result of the impact of WSSP on supply reliability. This trend has been validated through periodic customer interactions during the barazas. This also improved customers' willingness to pay, which has helped to increase revenue collection efficiencies to over 98%.

2.1.5 Prepaid metering

Prepaid metering also yields operational excellence in many ways. It reduces predatory effects of the middleman (the vendor) who obtains water from a water utility and resells it expensively (sometimes 10 times more) compared to the utility tariff. With prepaid meters on public water points citizens are enabled to purchase consumption credits at utility tariff rates. Prepaid meters on public water points were installed in slum areas of Kampala City prior to 2011. However, there was no corresponding human resource capacity building to cope with the operational and maintenance challenges of the technology. Emphasis had, initially, been put on social and community marketing and by 2013 it was clear that the technical challenges were destined to negatively impact the sustainability of the investment. There were, therefore, deliberate efforts undertaken by the new management to address this dire situation. The technical side has been strengthened by putting in place proper staffing, training and deliberate restocking of spare parts for maintenance. In addition, the reporting lines for the unit managing this technology were streamlined and managerial incentive plans put in place to enhance productivity. The management unit also looks beyond donor investments to increase penetration using internally generated funds. In the communities served with prepaid meters, it is clear that the people like the technology.

The other category of customers served with prepaid meters is the government institutional category. The new management, working with the guidance of the Ministry of Finance, Planning and Economic Development installed prepaid meters for strategic government institutional accounts where payments were regularly not being made: funds were being received from the Government and diverted to other purposes, on the pretext that there was no urgency in paying water bills.

Figure 2.8 Dr Eng Silver Mugisha (Third Left) with the Kampala Water Management Team inspecting the performance of the prepaid water meter in the informal settlement of Bwaise Kimombasa in Kampala, April 2017.

Locations where this occurred included police, prisons, defence, hospitals, and the Ministry of Finance, Planning and Economic Development. The main rationale behind the change was to ensure that these accounts do not accumulate arrears. While enforcement of credit limits has been quite a challenge because of budget constraints in the Ministries concerned, the move brought discipline and awareness.

2.1.6 Strengthening water quality management

Water quality management has been part and parcel of NWSC operations for more than 20 years. However, over time, considering investments, the frequency of restocking reagents and vital equipment, reskilling staff (leave alone restructuring to match emerging scope and challenges) and incentivising staff, the level of emphasis on water quality management was getting worse, year in, year out. By 2013, the state of water quality management was at its lowest, characterised by low staff morale, inadequacy of equipment and of chemical reagents, and inappropriate organisational structure. The effect of all this was unpredictable water quality, with instances of poor quality manifesting at customers' premises. Staff had also resorted to rent-seeking through private work while using

Figure 2.9 Joint verification exercise of the water quality tests in Kampala between NWSC water quality staff and team from CDC at NWSC Central Laboratory in Kampala, May 2015.

the corporation facilities. Staff appraisals and promotions were becoming unheard of, the operation was being seen as a 'prison' and occasionally regarded as a dumping ground for unwanted staff in the corporation.

The new management in 2013 could not let this appalling situation continue! In line with the new Strategic Direction, the department was restructured and new positions were filled with new staff and managers who were more qualified and committed. An internal audit identified staff with poor organisational behaviour and questionable integrity, who were then relieved of their duties and replaced. There was also a review of logistics requirements and a budget was apportioned to restock the department through a deliberate affirmative action. When staff witnessed the arrival of new equipment and chemical reagents, it caused a collective sigh of relief (and renewed motivation) that management had, at last, directed equitable attention to a department that was so critical to the delivery of good quality water services. More modern equipment has also been delivered through measures associated with donor-funded projects that have been implemented in the corporation since 2013. It is therefore not surprising that management is not afraid to collaborate with any outside accredited laboratory to check the legitimacy of periodic internal tests. This is because the staff and equipment of NWSC are, in most cases, superior to outside resources as a result of the affirmative investments undertaken. An example was when the water quality of NWSC was queried by the US Centre for Disease Control (CDC) in 2014, claiming there was contamination that was causing typhoid in the country. A joint verification exercise between NWSC and CDC staff revealed that samples had originally been taken outside the NWSC system and that the NWSC laboratory, sampling procedures and test kits met international standards. This was after independent tests by the Uganda National Bureau of Standards (UNBS) had given NWSC a clean sheet of total compliance. That explains why the NWSC laboratory has taken bold steps towards international accreditation to derive the full benefits of customer trust.

In order to strengthen the effective water quality regulatory environment, NWSC works collaboratively with independent laboratories both inside and outside the country. The Directorate of Water Resources Department of the Ministry of Water and Environment is the official water quality regulator, while the Uganda National Bureau of Standards has a statutory duty to ensure that all products supplied to Ugandan citizens conform to National Standards. There are also other private internationally recognised laboratories in Uganda that conduct specialised laboratory tests, such as like testing for heavy metals, which are sometimes used (at a cost). In all these cases, the work of water quality testing is collaborative to minimise conflicts arising out of sampling methods and locations. Results from these collaborative tests are regularly publicised to build customer confidence in the services that NWSC offers.

2.2 PROCESS INNOVATION

The other area that the new management undertook to improve (to buttress envisaged NPM reforms in the Strategic Direction) was process innovation. In doing so, the new management was externally open: learning from the prevailing environment of the corporation, acknowledging the need for major changes and ready to depart from previous routines. In other words, management was guided by the need to move forward into uncertainty, continuous learning, exploring new opportunities, feeling a sense of participative contribution and triggering creative discussions. Consequently, a number of transformative inputs and tools were created, as described below.

2.2.1 Revitalising tendering processes

The tendering approach for acquisition of works, goods and services had historically become heavily centralised at headquarters. This was not only unnecessary bureaucracy but was also disempowering managers in operational areas, costly and hampering infrastructure growth. There was really no value added in centralising most of the tendering processes at headquarters. Moreover, with increased scale of operations envisioned in the new Strategic Direction 2013–2018, centralisation was – sooner rather than later – going to be a self-defeating strategy. Of course, it was also limiting in terms of managerial innovation and accountability of managers whose task it was to make things happen on the ground. What was cardinal, though, was to strengthen the control environment so that risks of abuse of office and resource allocation inefficiencies were minimised.

Consequently, tendering processes were restructured, allowing managers to handle higher financial thresholds and become more innovative. However, this was done in line with the provisions of the country's procurement laws. In addition, the monitoring routines were strengthened, requiring more periodic quality reports to make sure nothing went wrong. At the same time, more time was devoted to capacity building of those handling tendering processes through on-job training, coaching and mentoring. As a result, managers became more innovative by introducing smart partnerships with local communities where the community would excavate trenches (with specifications provided by NWSC) and the corporation then delivered and laid the pipes. This helped to lower the cost of investments and fast-tracking implementation. New managerial innovations would easily be relayed to different areas for possible implementation through benchmarking, facilitated by regional managers. Most of these innovations would be discovered through daily engagement on regional WhatsApp platforms and/or through regular meetings of regional managers and top management.

In order to bring down the cost of investments further, the tendering processes were further modified to allow for a 'force account mechanism' where individual construction inputs are tendered for differently, maximising value for money. The labour component is paid for directly by designated corporation staff by engaging direct labour and providing accountability. This approach has helped to more than halve the cost of investment in comparison with if the entire works were put out to tender using the usual FIDIC tendering procedures. In addition, the new management has put more emphasis on use of internal capacity to carry out engineering designs, where there is moderately low technical complexity, which has helped to reduce investment costs further, while maintaining construction product quality. Moreover, working closely with the communities, local political leadership and incorporating a strong social marketing component has helped to yield a higher sense of ownership and costs such as compensation for land, especially where pipes are laid underground for wider public good has occasionally been waived.

At a strategic level, the Board of Directors also gave more powers to the accounting officer to undertake tendering processes, in line with the law, without referral to the Board. In this regard, the approval financial thresholds were raised from Ush 30 million to Ush 200 million to lessen bureaucracy in tendering processes. As pointed out before, this explains why the rate of pipe network construction has increased from 80 km per year in 2013 to over 1000 km per year during the period 2013–2017. The Board of Directors, however, have maintained their oversight role by reviewing and raising emerging issues (if any), based on mandatory quarterly reports on all procurements undertaken.

2.2.2 Towards universal water service coverage

Water utilities in developing countries continue to grapple with the issue of universal water access for citizens. In the quest to address this quagmire, the Board and Management of NWSC came up with an innovative universal water access project, code-named '100% Service Coverage Acceleration Project' (SCAP100). The key features of the three-year SCAP100 include a funding requirement of Ush 214 billion (about US$ 60 M), of which 40% is to be provided by the Government of Uganda through the consolidated fund and 60% from NWSC's internally generated resources. The project is incentivised by the 'water for all' mantra, with targets to construct 8000 km of new pipe extensions, achieve 140,000 new connection hook-ups, reach 12,000 additional villages and establish 20,000 new public water points. Each public water point serves about 20 homesteads, which is estimated at 200 people within a walking distance of 500 m. In addition, the project embodies a strong focus on investment cost containment through direct labour contracting and community participation. Figure 2.10 shows the roles and incentives of key stakeholders involved in the implementation of SCAP100.

SCAP100 has a number of implementation challenges. The target population has become moving target because of the immense appetite for

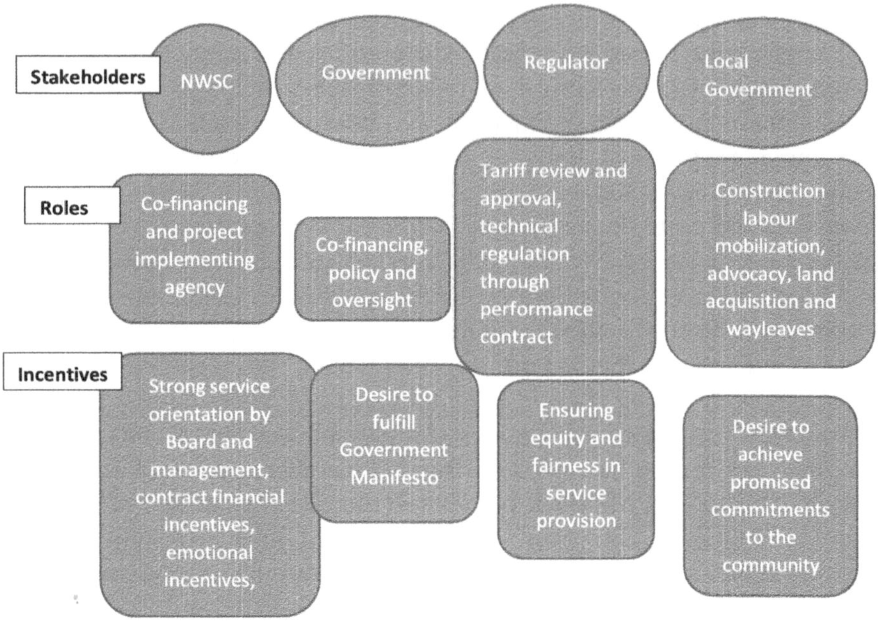

Figure 2.10 Roles and incentives of key stakeholders.

Figure 2.11 H.E the President of Uganda Yoweri Kaguta Museveni (third right) hands over a bucket of water to the resident of Katunda Village in Bushenyi as part of the launch Public Water Point in Bushenyi Area. August 2014.

Figure 2.12 H.E the President of Uganda Yoweri Kaguta Museveni handover a bucket of water to the resident of Dokoro as part of the launch of Public Water Point in Dokoro Area, March 2017.

NWSC services and perceptions of increased public accountability. The finance from payment of government water bills does not come in a timely manner. There is continuous political pressure to increase geographical coverage and peoples' service expectations are ever increasing (expecting a service today and not tomorrow). Furthermore, climatic change effects call for expensive resilient measures to mitigate demand outstripping water supply. However, these challenges are not insurmountable. There is significant Government support to SCAP100 in terms of budget provision and enabling policy. Also, local government provides advocacy and mobilisation of local labour. The existence of in-country pipe-manufacturing capacity also provides efficient project logistics supplies.

2.2.3 Capital projects implementation

The implementation of capital projects (small, medium and large) was another area that the new management had to tackle as part of the revitalised infrastructure policy in the Strategic Direction. As already pointed out above, most of the designs for medium to large capital projects were, prior to 2013, carried out by outsourced consultants, albeit with the presence of internal technical capacity. This was a huge unnecessary cost burden to the corporation that necessitated an ambitious rationalisation strategy. The use of corporation engineers for small and medium-sized projects had to become standard practice. The advantage of using a mixed approach – medium-sized projects designed internally and large (mostly donor-funded) projects by outside consultants – was that the corporation engineers felt challenged and got a strong sense of satisfaction through job enrichment. As a result, their technical design skills have been strengthened and careers enhanced. In reality, it is not only engineers on the internal design teams: there are also surveyors, social workers, economists and financial experts.

Apart from using a mixed design approach, public accountability in implementation of capital projects was strengthened in a number of ways. Beginning in 2013, all capital projects were structured into annual plans code-named Infrastructure Service Delivery Plans (ISDPs) showing the area, detailed locations, pipe size and lengths to be laid, and expected time of implementation. The ISDPs were then published in all major print media for public information and at the same time to act as frameworks for the public to demand equitable investments distribution and implementation. The ISDP publication was also posted on an official NWSC website as a permanent record of reference. This public information dissemination strategy also helped to build a strong nexus between geopolitical interests and corporation work, in a transparent manner. When politicians realised that the corporation was endeavouring to show accountability, they made it much easier for NWSC managers to lobby for increased government support in payment of water bills and provision of investment subsidies.

The other element introduced in capital project implementation, especially for donor-funded projects, was extensive use of local area managers/engineers (the ones usually in charge of operation and maintenance in areas) as Resident Project Managers (RPMs). The use of RPMs, rather than previous practice of only using project managers from Headquarters, brought a lot of ownership of the work and a reduction of the blame game that used to arise when projects progressed to their operation and maintenance phase. The RPMs were given significant contract management powers (relating to route realignments, compensation, stakeholder management, among others). As a result, the RPMs have used their informational superiority for local conditions to quicken project implementation. This use of RPMs has also helped to

Figure 2.13 Areal view of the Lubigi wastewater treatment plant in Kampala that was commissioned in 2014 and benefited from the QAC. December 2016.

reduce the project administrative costs that came with extensive travel from headquarters to project areas to solve simple contract management issues. The construction snags during defects liability period are also more efficiently handled.

Figure 2.14 Areal view of the Bugolobi Wastewater Treatment Plant in Kampala under construction that benefited from the QAC. January 2018.

In order to manage contract variations and enhance value for money, a quality assurance committee (QAC), comprised of senior engineers outside the project management teams, was also put in place. The QAC reviews and approves all variations from contractors and consultants, which have been received and recommended by the project managers. This was deemed a very important step before the proposal is sent to the contract committee for award of contract amendment. The QAC step was necessitated by the fact that the contracts committee does not normally have sufficient time to probe all the technical aspects to establish value for money and was becoming a rubber-stamping unit. The QAC also has the mandate to demand periodic reports from the consultants, through the project managers, to rule out the possibility of professional laxities that can result in time and cost overruns. If the latter is not strictly monitored, consultants can fleece clients through costly time extensions.

2.2.4 Performance improvement planning

Prior to 2013, the linkage between strategic and tactical planning was weak. There was a corporate plan, but it had a weak linkage with annual action planning. There was therefore a need to have a stronger linkage between the two levels of planning. After the promulgation of the Strategic Direction 2013–2018 and the Corporation Plan 2015–2018, it became very clear that their effective implementation and the eventual realisation of objectives would be heavily dependent on well structured tactical action plans (TAPs). Moreover, this was also not enough: there was a need for closer working together of corporate planning, monitoring and evaluation (M&E) and operational staff in regularly conducting monitoring and evaluation. This had to be coupled with regular feedback to operational staff on progress of delivery of intended results. At the operational level, meetings were conducted based on observed performance results of the immediate past period. For example, revenue billing, collections, arrears recovery, water sales, water production, expenditure, reduction of illegal connections, connection of new customers had to be monitored on a daily, weekly and monthly basis to avoid performance gaps relative to targets. In this manner, staff meetings became more results oriented.

In addition, TAPs had to be regularly informed, reviewed and updated using findings and observations from regular operational visits from headquarters by the Managing Director, Deputy Managing Directors and other Directors. The main consideration and basis of the changes would be emerging overarching policy directives, changed business environment, new customer expectations and the observed extent of achievement of strategic and operational targets. This approach has made a fundamental difference from the past because it has helped to keep staff more engaged and the TAPs more live and adaptable. As a result, it has been much easier to roll over TAPs across consecutive years, based on more accurate data and relevant external operating context. Furthermore, because local political leaders and communities are invited during the above visits and asked to give their inputs, the TAPs remain both internally and externally authentic and relevant.

TAPs have also regularly received a 'face-lift' from periodic analysis of commentaries and observations by customers and other stakeholders via regular staff call-outs/surveys, social media information, courtesy calls and direct phone-in by outsiders. In the past, attention to this perspective was structurally limited, given that it was only the Managing Director's concern. This time round, the approach has been mainstreamed into the managerial structures of the corporation so that action and feedback can be handled in a collective and more efficient manner. The results of these interactions are also discussed in staff meetings and corrective action agreed upon. Consequently, this has become a value-adding managerial practice, which has helped to strengthen customer care in the organisation. Exhibits #11–12 give some examples of facets of performance planning and outcomes.

EXHIBIT #11: FROM SAM APEDEL (2018), NWSC PR MANAGER

Clear and transformative mind-set of the MD was key. The visionary and bold leadership with a can-do attitude, for example, the expansion from 24 to 225 towns is majorly attributed to that ambitious *'we have to do it'* mind-set. The MD inspired management to ignore all the doomsday predictions and march forward with unprecedented geographical expansion. Hence, management took calculated risks and refused to be tied by straightjackets of budgets and bureaucratic procedures, while focusing instead on outcome and delivering water for all. In addition, there were vigorous strategic alliances and building of bankable goodwill with the President, Ministry of Finance (MoF), Parliament and Communities.

EXHIBIT #12: FROM GODFREY MWESIGYE (2018), NWSC BRANCH MANAGER

Over the last four years we have experienced growing high level of NWSC's acceptability by communities. Customer satisfaction index has gone high in comparison to the days of takeover of these new towns. In addition, NWSC has built goodwill in comparison with other utility sectors or other Government agencies since 2013. We have seen significant staff development, training and mentoring (organisational fit) from local councils, local government. Furthermore, there has been increased staff productivity and reliability. The staff complement has gone up. Integrity of staff is something we are proud of (despite a few cases here and there). Our performance is second to none in service sector and other entities. Customer complaints resolution time has improved. In-house IT system development (like billing interface, e-procurement and many more). Today, I received a call from the District Chairman, later from the Chief Administrative Officer in appreciation of NWSC good works in the Paidha and Nyapea, mostly via the 100% Service Coverage Acceleration (SCAP100) project. The Resident District Commissioner has been giving me airtime to go on a radio talk show, in appreciation. The decentralisation of management has been a great performance driver. Managers are empowered, targets attached to them, action plans developed – creating a sense of business ownership. Regular performance evaluation is a unique strategy in Uganda: NWSC has become a benchmarking entity for various government sectors.

2.2.5 Improving communication processes

Communication and publication relations has increasingly become a key ingredient in sustainable utility operations management. Organisations that do not manage this aspect well will achieve sub-optimal results. When the new management took over in NWSC in 2013, there were structural gaps in communication and public relations management. The gaps were in both internal and external dimensions. Internally, there was inadequate staff engagement due the manual nature by which the function was mainstreamed in the organisation: staff engagement was predominantly through executive visits, which were also irregular and unpredictable. Whenever visits happened, they lacked the emotional intelligence attributes of self-awareness, self-regulation and empathy: instead of imparting positive emotional momentum, they would sometimes leave behind lasting negative impacts by creating fear and employee demotivation. The reverse was also true: some managers were visiting operational units purely with a view to earning per-diems and giving feedback that was characterised by telling staff what they wanted to hear. Of course, this was not productive, much as it did not impart negative emotions. Exhibit #13 contains some perceptions about management efforts to enhance communication processes.

EXHIBIT #13: FROM STEPHEN WALAKIRA (2018), NWSC AREA MANAGER

In the last four years, a lot has been done to reduce communication gaps in change management. When change is announced, staff have information concerns. Often, leaders want to explain why the organisation is moving in a certain direction and why the change is a good idea. This was a challenge before. Staff normally don't want to be told the change is good until they understand it. Leadership has been sharing information as plainly and as completely as possible to the lowest ranked staff through different communication platforms like social media and new journals among others. In the presence of clear, factual communication, staff disseminate clear information about corporation changes, and have minimised rumour mongering and corridor lobbying manifestations. Leadership has tried to respond to staff/stakeholder concerns/questions and created a sense of ownership among the general employees and public.

Consequently, the new management had to restructure the entire communication and public relations (C&P) strategy of the corporation to ensure effective information flow. The C&P department had to undergo major restructuring to give it the prominence and momentum it deserved. Originally, there were only two staff in the department, which was making it overly ineffective and inefficient. The staff numbers were increased, new functional units created and necessary mergers carried out. In this regard, and for purposes relating to ***strengthening external communication***, the call centre unit was restructured, enlarged by benchmarking best practice in internationally well performing utilities (in the Netherlands and South Africa) and put under the C&P Department.

Figure 2.15 The NWSC call center inbound staff attending to the customer calls at the NWSC Call Center Bugolobi Kampala. September 2016.

The operations of the call centre had to be enhanced through core call agents for inbound and outbound calls; regular surveys on feedback introduced and emotional incentives for well performing agents were put in place. The status and remuneration structure of call agents was also improved to increase retention and a better sense of belonging.

Furthermore, a new unit to handle social medial issues (Facebook, Twitter, LinkedIn, and the website) was created. The new Managing Director was also given a personalised Twitter handle to help regulate and reinforce that aspect of social media. A benchmark for response rate of not more than 20 minutes was specified, within which the social media team needed to respond to customer concerns. The Managing Director checks adherence to this benchmark on an unannounced basis. The social media team has put in place innovative informational sensors and networks to detect any adverse information about NWSC and provide remedial measures. This approach has created customer confidence in social media mode of communication and eventually helped in building a strong corporate brand and public accountability. Exhibits #14–15 contain some examples of stakeholder engagements and underpinning activities.

The other aspect of C&P management improvement was creation of the Corporate Relations Unit (CRU), directly under the Managing Director's

EXHIBIT #14: FROM SAM APEDEL (2018), NWSC PR MANAGER

Deeper stakeholder engagement and embracing of traditional media and social media for outreach and engagement has tremendously improved communication processes of NWSC.

EXHIBIT #15: FROM DANSTAN PEKO (2018), NWSC BRANCH MANAGER

We have observed significant improvement in performance of NWSC. This is attributed to a number of innovative activities. All these have been made possible due to good stewardship of the corporation by the board, our dear Managing Director and the executive management of NWSC, who were able to champion performance and rally employees with good attitude to better performance at all levels. Those with poor attitude have been continuously mentored and coached. Also effective engagement of stakeholders at various levels has contributed greatly. For example, at the branch level, we have used local leaders, congregations at churches/mosques, phone messages and, of course, not forgetting teams at the Centre who handle management of social media platforms and the Call Centre.

office but with a functional linkage. The CRU handles the personal assistance functions of the Managing Director, include logistical office support services and stakeholder relations management. The unit carries out daily surveys of key operational activities to get customer feedback, receive staff integrity information from stakeholders, and prepare periodic briefs to the Managing Director. Through this approach, the traditional secretarial administration function has been significantly transformed in terms of productive value. The CRU is also one of the sources of real-time staff emotional perceptions (about their well-being and concerns) to the Managing Director. The information is, however, cross-checked with other sources for validation, accuracy and authenticity.

Continuing on external communication, other channels were introduced. These included regular information, education and communication (IEC) activities by NWSC managers and staff in religious institutions, markets

Figure 2.16 Dr Eng Silver Mugisha (forth right) sensitizing congregation on NWSC operations during a church service in Mbarara. May 2015.

Figure 2.17 Dr Eng Silver Mugisha (left) sensitizing the community after commissioning water supply project in Kanungu one of the operational areas of NWSC. March 2017.

Figure 2.18 NWSC staff (right) sensitizing and getting feedback from the customer in Owino Market in Kampala. November 2016.

and other community gatherings. The message of IEC in these channels included: tariff charging systems, new development programmes, completed projects, requests to the clients and expected staff conduct. These IEC platforms also help as feedback-giving opportunities for customers. This IEC managerial approach has greatly helped to inspire practical behavioural change among NWSC managers and staff, which has in turn strengthened the 'customer is the reason we exist' organisational culture.

On the ***internal communication*** dimension, the new management introduced a plethora of WhatsApp groups, covering all operational segments of the same nature. It is on these platforms that all events happening in different operational units are posted for information, action and benchmarking. A similar system has also been established for customers in specific local contexts for tailor-made IEC activities. Another internal outlet that has been introduced is a Daily Staff Newsletter (DSN), which captures significant daily events that should be known by all staff in the corporation. The events normally include information relayed through visits by the Board and/or Managing Director, integrity issues handled and matters of celebration. The DSN is normally in a mixture of pictorial presentations, captions, brief narratives and intelligent quotes. There are also other regular announcements relayed through the email systems, like daily comparative revenue billing and collection figures, monthly performance and awards, death announcements and internal memo communication of diverse nature and purpose. On top of this, the Managing Director occasionally sends audio voice recordings (through WhatsApp), mainly on major performance achievements, behaviours that require to be bounded and future plans. The timing of these messages is sometimes targeted for major calendar days, especially selected public holidays like Religious Days, Independence Day, Women's Day and Labour Day. Other communication channels include: regular staff social get-togethers (like every last Friday of the month), targeted lunch/dinner, sports events and other meetings. This approach has helped to bring better staff cohesion, teamwork and a stronger sense of belonging.

2.2.6 Information technology (IT) systems

Another area of innovation by the new management that helped to add value to utility operations management is information technology (IT), specifically applications development. A lot of work had been done in the past but more was needed to spur further technological change and eventual increased productivity. In fact, much effort had been applied in developing automated financial management systems, billing, bank payments and procurement. However, the billing system was based on an outsourced patented software application from South Africa and the corporation was incurring about US$ 300,000 per year in licence fees on top of costs of maintenance and troubleshooting. This cost was expected to increase significantly with the increased number of users resulting from increased geographical footprint. Other systems apart from the e-procurement system also had huge licence fees and maintenance costs. Hence the cost of doing business was destined to skyrocket, given the ambitious expansion programme in the Strategic Direction 2013–2018. Consequently, the new management came up with strategic interventions in the area of applications development to cope with this challenge.

In 2014 a new billing system that was designed and developed by the corporation's IT staff was implemented, replacing a South African originated system (CUSTIMA), which had been in operation for over 20 years. The newly implemented system was web-based, in contrast to the previous system that had to be installed individually in every additional operational business unit, through an independent server system. The new billing system also had more features and applications, including the customer relationship module (CRM) and a new connections module. In addition, it was easy to integrate it with the newly introduced e-payment system for easy revenue management and monitoring. The implementation of a new billing system in the corporation has brought many operational advantages: it has eliminated the need for payment of annual licence fees for existing connections (over US$ 300,000 per year). In addition, system modification, integration and troubleshooting are much easier since it is internally designed. Moreover, the fact that it is centrally monitored through a central server system means that all forms of cybercrime and fraud risks have been minimised. As a result, it has been easy to increase the geographical footprint of NWSC from 23 to 218 towns without additional fixed costs as was the case with CUSTIMA.

As mentioned above, another important new system is the customer relationship module (CRM). The system, which is an integral part of the new billing system, has helped to improve management of customer care operations. All areas and branches are required to input all reported customer complaints into the CRM. The system then helps to return all complaints registered, resolved and those pending action. The time taken during each of the stages is also captured for effective monitoring and evaluation. A core unit of staff is charged with daily monitoring and demanding feedback and action from operational units. All senior managers are then presented with a daily performance scorecard of areas/branches for information and appropriate action as required. The introduction and application of CRM has greatly improved customer satisfaction and impacted positively on the corporation's public accountability efforts. As part of integration with the billing system, staff have also developed an e-invoicing system for water customers through use of (rugged) smart phones which take meter readings on site and relay them through a seamless SMS system to the e-billing system. The system then generates a provisional invoice that is presented on the spot to the customer as a payment notice, pending a permanent invoice. In some cases bills are sent via e-mail through the same system. In the same connection, an NWSC app has also been developed, based on the Android platform, which is accessible to all customers, using individual unique customer references. The app can be used to access outstanding amounts and payments. It can also assist in sending queries to the corporation management.

Other systems that have been developed to improve administrative efficiency include: e-inventory for efficient logistics management and e-leave application for efficient management of staff official leave. Just

Figure 2.19 A screenshot of the NWSC mobile app developed by the NWSC applications development team. February 2018.

like the e-procurement system that was developed around 2009, these systems have different approval levels that are password protected to avoid manipulation or administrative failures. The approval levels have been set in line with the existing policy guidelines of the corporation approved by management and Board. The systems have tracking systems that are easily auditable to minimise risk exposure. In the same vein, an e-invoicing and payment system has introduced to monitor external water quality services provided to external clients. Previously, there was much fraud in this area, with significant rent-seeking by staff as a result of the weak systems in place.

2.2.7 Improving risk management

Risk management is increasingly becoming a central part of good corporate governance for any organisation's strategic management. It is a process whereby organisations systematically address the risks relating to their strategic activities with the goal of achieving sustained benefit with each activity and across all activities. The focus of good risk management is the identification and treatment of such risks. Its objective is to derive maximum sustainable value from all activities of the organisation.

Effective risk management depends on a strong understanding of the potential upside and downside of all the factors that can affect optimal organisational business conduct. It increases the probability of success and reduces both the probability of failure and the uncertainty of achieving the overall organisation's strategic and operational objectives. It therefore goes without saying that risk management should be a continuous and developing process that streams throughout the organisation's strategy design and implementation. It must reflect on the past, present and most importantly the future of the organisation.

Risk management was introduced in the corporation in 2001. It was mainstreamed through the internal audit function, requiring that all internal audit findings and determinations had to be calibrated by the level of associated risk (either low, medium or high). This simple approach to risk management was in place up to 2013. The main upside to this approach was that it was simple to apply; however, there were a number of downsides. It was only from the audit perspective and was not mainstreamed at strategy design stage. Consequently, the approach was reactive in nature and heavily deficient in preventive aspects. Risks were only identified when they had already occurred, and were merely characterised and categorised for corrective action. Moreover, the calibration process of these risks was highly subjective and lacked scientific input to increase their objectivity. As a result, their identification and inclusion in audit reports and findings tended to be merely a ritual for appeasing stakeholders, especially external corporate governance enthusiasts. Other corporation managers and staff did not take risk identification seriously. Consequently, the exercise remained well below the threshold for achieving organisational objectives and goals.

The new management, as part of effective implementation of the Strategic Direction 2013–2018, incorporated a paradigm shift to risk management. Instead of waiting for risks to occur, as was the case previously, risks were identified across all the strategic priority areas (SPAs) of the strategic direction. The identification process was undertaken by a multi-disciplinary team of corporation managers with a diverse understanding of operations. Risks were then themed, coded and prioritised. As a result, a risk register was prepared with top priority risks in each thematic area. The areas included technical, financial, commercial and human resources management risks. Teams were then constituted to ensure effective implementation of the risk register. It should be noted that the risk register delineated arrays of risk matrices with well quantified probabilities of occurrence, remediation measures and cost implications. In a related development, a dedicated risk management department was created and staffed with well qualified staff and managers with good skills and experience in risk management. In addition, a separate risk committee of the Board was created to give prominence and oversight to the risk management team in the corporation. Consequently, risk management became better mainstreamed in the corporation, which in turn increased predictability and sustainable achievement of organisational strategic objectives.

2.2.8 Other initiatives

There are a number of other process innovations put in place by management to bring marked technical change in operational processes. A new concept called ***Know Your Customer*** (KYC) was designed and implemented around the end of 2014. KYC enables every customer invoice to have a printed telephone contact of a specific staff member who can be contacted directly to offer quick customer service. The approach personalises customer handling and removes bureaucracies that plagued customer care operations before. The staff members assigned are senior in the organisation and have good customer care skills. This strategy has enhanced the corporation's service offer and hence improved customer delight and trust.

The other initiative was the introduction of an ***energy reduction strategy*** in water treatment plants to contain the costs of production. The strategy involved the corporation's engineers using a multiplicity of skills to conduct systematic feasibility and research studies. The study process was highly participative and collaborative to maximise generation of ideas and solutions. A comprehensive strategy was then presented to management, with proposed budget estimates for implementation. Briefly the strategy involved installation of power factor correction devices, replacement of low-efficiency pumps, optimisation of production processes and a number of other hydraulic modifications. Management put in place an incentive mechanism tagged to energy reduction targets. Within a period of one year, some water treatment plants were able to register up to 15% reductions in energy costs. In fact, a large proportion of cost inputs for energy reduction was eventually financed through cost savings. In some cases, donor agencies like GIZ of Germany provided seed financing and teamed up with equipment suppliers who were eventually paid through cost efficiency gains realised by the corporation.

In order to increase climate-change resilience, the new management also came up with alternative water supply measures involving ***use of underground water sources***. It must be noted that most of the water production systems of the corporation depend on surface water sources. With increasing occurrence of extreme dry weather events, surface water sources were under growing pressure from rising demand patterns. It was therefore deemed necessary to build climate-change resilience measures through motorised borehole systems. Of course, this had to go through the entire process of hydrological resource mapping, drilling, pump tests and installation works where the amounts of water available were found to be adequate. In all cases, water has to be pumped to a raised reservoir facility and distributed to customers by gravity. This option has greatly helped to stabilise water supply and to reduce the costs of laying long pipes from distant sources.

2.3 STRATEGIC ALLIANCES AND SOCIAL RESPONSIVENESS

Strategic alliances and social responsiveness are also vital elements of the internal perspective of any organisation. Up to 2013, there were a number of ways in which areas would collaborate internally to exchange ideas and reinforce innovation potential. However, there were a number of other ways that had to be introduced to strengthen this approach.

2.3.1 School water and sanitation clubs

There are a number of clubs that provide social engagement for schoolchildren. These include debating, music, dance and drama, scouting and sports. The corporation's Strategic Direction 2013–2018 put particular emphasis on environmental conservation. However, this was not a strategic task that the corporation could implement on its own, given that it is not its core mandate and yet lack of it affects its operations directly. The corporation chose to make its contribution through tree planting, especially in the catchment areas of its water resources. Schoolchildren, given their large numbers, were seen as potential partners in this noble task. At the same time, the new management was aware of the need to promote efficient water use and hygiene management in schools, especially among the young children. This was because the ultimate impact of water supply and sewerage services would not be optimally achieved if this gap was left unattended. Consequently, in 2014 the corporation introduced School Water and Sanitation (SWAS) clubs in

Figure 2.20 The SWAS Patron – Ms Sylvia Tumuhirwe (first right), NWSC Managing Director – Dr Eng. Silver Mugisha (second right) and NWSC Board Chairman – Dr Eng. Christopher Ebal (third right) join the winners of the SWAS Convention 2016 under the theme "plant a tree save the future" at the NWSC-IREC Kampala. July 2016.

Figure 2.21 The pupils of Greenhill Academy exhibit their SWAS project during the SWAS Convention 2016 NWSC" at the NWSC-IREC Kampala. July 2016.

schools. The cardinal objectives of SWAS clubs hinged on promotion of hygiene through proper sanitation and water use (hand washing), efficient water use, and promulgation of sustainable environmental conservation practices.

The SWAS clubs were introduced in a phased manner because the project required significant advocacy, incentive application, resilience and change management. Different school administrations were visited at different times by the corporation's core team of SWAS champions, who explained the concept and objectives of SWAS clubs, arguing for each school to allow their children to join. The corporation's champions went with a full package including printed strategic outline of SWAS clubs and a promise of seed start-up funds if the school administrators embraced the SWAS idea and began implementing it. The seed funds would come from the corporation's corporate social responsibility (CSR) budget. Most school administrators were quick to embrace the SWAS club idea because of the clarity of its objectives, the value proposition it was bringing and because it offered an opportunity to partner with NWSC. Within three months, over 30 schools, at primary and secondary levels, had signed off memoranda of understanding (MoU) with NWSC to implement SWAS clubs. As a start-up activity, each of the first schools were organised into a simple management structure comprising a patron, who was a teacher, and an SWAS coordinator, who was a student. The rest of the interested children were then organised under these two heads. On the other hand, the corporation's champion team was also headed by an overall SWAS club patron. To facilitate the smooth running of the SWAS function, the corporation also recruited a liaison person, on contract, to carry out day-to-day coordination. The marketing of the clubs was so thorough that within a short time, clubs would get hundreds of students/pupils joining. It should be noted that the SWAS club campaign started in schools around Kampala City for ease of communication and management of transaction costs involved. After the idea was fully piloted and entrenched, it was rolled out to upcountry towns where the corporation operates.

The main groundbreaking SWAS club activity that challenged more schools to join was the June 2014 SWAS club convention that was held at NWSC's International Resource Centre (IREC) premises. It was a joint effort of the corporation's team of core champions and representatives from SWAS club schools. The function was presided over by three Ministers (Water and Environment, Health, and Education). It was a bee-hive of activities, comprising an exhibition of innovations, drama, debates, music, poems and skits in the areas of sanitation, efficient water use and environmental protection. These were organised in a competitive environment with professional assessors who helped to choose winning schools in each category. The winners were rewarded with cash prizes and physical presents. The patrons and coordinators of the top performing schools were awarded fully paid trips to France for benchmarking best practices in water, sanitation and environment management. All these activities were covered by both electronic and print media for wide publicity. Subsequently, the SWAS club spirit spread like a bush fire through all schools, which resulted in more enrolment and increased participation in the following annual conventions. As of June 2017, there were over 150 SWAS clubs in the whole country and the number continues to grow, with sustained advocacy efforts.

The introduction and implementation of SWAS clubs has brought several benefits to NWSC. The clubs serve as good instruments to carry out environmental protection activities. This has been done through a number of ways. As of June 2017, more than 500 tree species had been planted by SWAS clubs to protect the environment. The clubs are also contributing to the planting of more than a million trees in different towns under NWSC jurisdiction. In addition, the SWAS club children, through drama, music, debates, skits and poems have increased awareness in their communities of water, sanitation and environmental conservation. Furthermore, the corporation's brand has been greatly enhanced by supporting SWAS club activities. The children, parents, relatives and indeed the general public have appreciated NWSC sustainability efforts through this type of giving-back activity. There is an SWAS club anthem that incorporates central aspects of NWSC values and strategic tenets, which when sung at different forums helps to promote the NWSC brand further. This has tended to increase awareness and public accountability.

2.3.2 Uganda young water professionals chapter

The Uganda Young Water Professionals (YWP) chapter was originally set up around 2008 under the auspices of International Water Association (IWA). However, it had been created with an unclear structure. The original intention was that it would operate on the East African regional level, but without a proper structure and objectives, the work of the Young Water Professionals lacked authenticity, grounding and accountability. In 2014, the new management of NWSC saw it fit to revive the YWP chapter of Uganda. A proper management structure was put in place and filled with young staff through a more transparent process. After filling the structure, recruitment of other members (who were aged 35 years and below) from within and outside the corporation was carried out. YWP was seen by management as an important tool to implement the environmental protection agenda that was enshrined in the Strategic Direction 2013–2018. To create momentum, the position of patron was created, to be occupied by a senior member of executive management of the corporation. In fact, the patron of the SWAS clubs was also asked to be the patron of the YWP since there was to be a nexus between the agendas of the two institutions. Accordingly, the YWP came up with an action plan that addressed three thematic areas: environmental protection, hygiene promotion (through proper sanitation) and efficient water use. The idea of a YWP action plan was a good one, especially after I became a Board Member of IWA in 2014. As a managing director of NWSC, I had to support implementation of the action plan as part of the broader objective of increasing visibility of Africa, starting from home (Uganda).

Like the SWAS clubs and sometimes working collaboratively, the Uganda YWP chapter has carried out a number of activities. They have championed tree planting in various areas of NWSC and beyond. Over

Figure 2.22 NWSC board and management pose for a group photo with the Young Water Professionals (YWPs) Uganda Chapter after the launch of the one million trees planting campaign in Ggaba Water Works Kampala. September 2017.

Figure 2.23 Dr Eng Silver Mugisha (Left) joins YWP-Uganda Chapter to plant trees at Ggaba Water Works Kampala as part of the launch of the One Million Tree planting campaign. September 2016.

one million trees have been planted over a period of three years since 2014. In addition, the YWPs have participated in car-washing activities, community sanitation improvement and running for social causes like cancer treatment. They have also joined the rest of the YWPs in summary reporting for conferences, especially those organised by IWA and the African Water Association (AfWA). Specifically, the Uganda YWP chapter has social media handles that are used to market their activities in Uganda. This has helped to increase the network of young upcoming water leaders, nationally and internationally.

The YWPs have helped to strengthen NWSC brand. Through their activities, the public has come to realise that the corporation has a deliberate strategy to grow the career of young people. Some of them who are leaders have found a strong sense of belonging and job satisfaction. Moreover, the corporate social responsibility (CSR) activities of the corporation have also been put in the spotlight by these enthusiastic YWPs. As a result, this has increased the public perception of NWSC's social responsiveness. Other young people feel like working with NWSC, partly because of the good work the YWPs are doing – in other words, the corporation is the one of the ultimate organisations to work for by most of the students that are finishing graduate studies.

2.3.3 Environmental conservation initiatives

As already pointed out, the core mandate of NWSC relates to provision of water and sewerage services in a financially sustainable manner. However, this mandate cannot be fully realised without due consideration for environmental conservation practices, especially for raw water sources where the corporation obtains its water for treatment. There are structures within the Ministry of Water and Environment that deal with this issue. However, with increasing pressure from climatic change effects, forest destruction and wetland encroachment, the workload on the Ministry structures has increasingly demanded collaboration from NWSC. Consequently, a number of collaborative initiatives have been undertaken to protect water sources.

One of the major initiatives has been massive tree planting campaign. As of December 2017, through collaboration with other stakeholders like politicians, schools and communities, NWSC had planted over 800,000 trees in the space of three years. The tree species planted were selected on the advice of professional institutions like the National Environment Management Authority (NEMA) and National Forestry Authority (NFA). The NFA has also invariably provided tree seedlings to boost the campaign. The targeted locations of tree planting have primarily been areas around NWSC raw sources to ensure effective catchment protection. The tree planting has been undertaken with a key objective of maximising socio-economic objectives in terms of reduced water treatment costs resulting from deteriorated raw water quality, preventing reduced water levels and obtaining social capital from neighbourhood communities. The socio-economic returns, of course, take time to fully materialise but they will surely come.

The other initiative has been massive community sensitisation and social marketing to reduce environmentally unfriendly practices. These include: source encroachment through farming activities, sand mining near river and lake intakes, source diversion for local agricultural activities and wetland drying through open channel drains. The NWSC approach was to work closely with local political leaders, sensitizing them to the dangers of environmental degradation affecting the water cycle so that they in turn can collaborate in sensitizing the local public. The main communication channels used have been radio and television stations, social media, churches and other strategic alliance meetings. This approach has registered significant success in terms of improved awareness, practical action to curtail bad practices, advocacy to budget for environmental

Figure 2.24 Dr Eng Silver Mugisha (second right) together with the Board members plant a tree as part of the One Million Tree Planting campaign at NWSC Kisoro Office.

Figure 2.25 Pupils and teachers of Gayaza Junior School Kampala plant 200 trees in Gayaza as part of the SWAS club drive to protect the environment. April 2017.

Figure 2.26 Young Water Professionals – Uganda Chapter planting trees along Entebbe Road Kampala as part of the drive to protect the Environment. August 2016.

protection activities at local government level and bottom-up lobbying for political support to deal with environmental degradation. No wonder for the whole of 2016 and 2017, His Excellency the President of Uganda was at the forefront of advocating for environmental protection at public gatherings and other focus group meetings.

In addition, NWSC has teamed up with other organisations like Kampala Capital City Authority (KCCA), NEMA, the Directorate of Water Resources of the Ministry of Water and Environment, and development partners (typically GIZ of Germany and World Bank), among others to carry out routine industrial pollution control. In this regard, regular pollution monitoring reports are produced and affected industry owners are asked to put in place remedial measures. One of the measures suggested to such industries has been the establishment of pre-treatment facilities to reduce influent loads to NWSC sewerage treatment infrastructure and thus to enhance process quality. Other measures have included stopping direct discharge to receiving water bodies to reduce environmental pollution of lakes and rivers. The collaborative arrangement is structured into strategic and operational levels to boost the effectiveness of decision making.

Another source of environmental degradation in urban settings where NWSC operates is lack of proper onsite sanitation, where conventional sewerage systems are lacking. The challenge has been tackled by NWSC in a number of ways. Using a combination of internally generated financial resources and donor funds, NWSC constructed over 300 onsite public toilet facilities all over the country (the majority in Kampala City) during the period 2011–2017. These public toilets have helped to enhance public hygiene management and environmental pollution control. The challenge with this type of sanitation technology relates to poor community management, occasionally, spurring conflicts among user families. NWSC has worked closely with the City, local town and village council authorities to resolve some of the emerging conflicts. In addition, NWSC has been active, working with local political leaders and technical staff, in carrying out targeted information, education and communication activities to encourage construction of private onsite toilets and/or septic tanks and ensuring their proper use, especially when such households have water connections. This has helped to reduce incidences of waterborne diseases, and hence improved health impacts.

Chapter 3

The people perspective

The other important business perspective that the new management considered with a view to transforming NWSC was the people perspective. In line with the Strategic Direction which had just been promulgated, the key elements of this perspective were: re-scoping of employee job definitions through restructuring, enhancing employee productivity, stepping up staff engagement practices, revitalising staff welfare, inculcating a continuous learning culture and a focus on emotional intelligence traits. Exhibit #16 gives some version of people perspective since 2013.

EXHIBIT #16: FROM CREAM NINSIIMA TREASURE (2018), NWSC ACCOUNTS OFFICER

It has been very impressive the way things have taken place in the corporation in the last four years in the area of human resource management. We have seen massive employee and stakeholder engagement; communication channels have been established to improve information flow (digital and non-digital); varied situational leadership styles exercised right from the Board to Executive management.

Author: Silver Mugisha
doi: 10.2166/9781789060256_0065

3.1 RE-SCOPING EMPLOYEE JOB DEFINITIONS

This element of the people perspective was addressed through a corporation-wide restructuring designed to organise the corporation's business in a way that would make it operate more efficiently in response to new and ambitious goals. Specifically, we sought to make NWSC operations run more smoothly and in an economically sound manner. The main rationale for including this in the Strategic Direction rested on the extensive business growth that had occurred over time (more than 10 years) without a matching organisational structure. As a result, most employees and managers had developed an intrinsic sense of resignation, with no meaningful possibilities for personal and career growth. This accumulated career stagnation, coupled with the new management's forward-looking growth possibilities, necessitated a major restructuring exercise. The corporation needed to restructure (internally) its operations, processes, departments and reporting structures to enable the business to become more integrated and profitable in the wake of envisaged rapid geographical expansion.

As already pointed out, the restructuring exercise was carried out through a protracted participative process, involving all internal departmental managers and union leadership. Job positions, descriptions, qualification and experience requirements were reviewed, modified and enhanced, including introduction of new positions. Three deputy managing director (DMD) positions (that of Finance and Corporate Strategy, Technical Services and Board & Management Services) were created. The DMDs were to head new organisational structures called business streams. Below the business streams, the new structure introduced directorates, headed by directors and followed by senior managers who were designated to head departments. The design approach that was applied was a mixed one, optimising flat and long organisational structural design. The bottom-line objective was to enable quick decision making while maintaining effective management control.

The other specific new functions that were introduced by the restructuring exercise included: business and scientific services, corporate events management, administration, office support services, investment and corporate strategy, programmes and performance management, regional structures, corporate relations management, digital marketing, integrated call centre management, corporate communications, vocational skills training, IT infrastructure and applications development. Most of these were introduced in response to emerging ambitious strategic direction aspirations and role ambiguity that was pertaining at that time. It was, therefore, crucial to create positions that would increase focus and commitment in key new areas that were necessary to support implementation of the strategic direction.

After management had come up with a draft structure, as was procedurally required, it was presented together with budget implications and implementation modalities to the Board of Directors for review and approval. The new structure was then approved with minor amendments. The Board directed that implementation of the structure be carried out in a phased manner, in line with emerging workload needs, staff productivity thresholds and affordability of the attendant restructuring costs. The Board also required that the implementation of the structure should be done in an orderly manner such that work was not disrupted. In addition, the principles of transparency, fairness, objectivity and integrity had to be upheld. The preferred approach to filling positions in the new structure, according to management recommendation and approval by the Board, was to declare all positions vacant and advertise them internally. Those who were occupying retained positions (other than new ones) were notified that priority would be given to them but they would have to demonstrate renewed drive and capacity through a competitive process. In case they were not successful, the Board would exercise its discretion regarding the next steps in line with the redundancy policies. This was an extremely delicate exercise, subjecting the entire top management (except the Managing Director) to tension with perceived fear of potential job loss! The adverts and shortlisting exercises were carried out by the Managing Director, assisted by a task force, comprised of some Human Resources Department staff and other staff not directly affected by the restructuring exercise. The selection interviews were conducted by the Board of Directors using competence-based criteria, which included knowledge, skills/practices and attitude. The latter mainly considered the character of the employee with regard to team-working, drive and energy, honesty, inspiration, forward-looking and alignment to the ambitious transformative focus enshrined in the strategic direction. The exercise was a resounding success, with all designated key positions filled by managers with renewed energy and commitment.

3.2 ENHANCING EMPLOYEE PRODUCTIVITY

The other key element of the people perspective which was incorporated in the Strategic Direction was related to enhancing employee productivity. This specific indicator refers to an assessment of the efficiency of a worker or a group of workers. Productivity may be assessed in terms of the output of an employee, given a specific set of inputs, in a specified time duration. In water utility industry, employee productivity is seldom measured in terms of the number of employees per 1000 connections, number of employees per unit of water supplied, number of employees per 1000 km of network length, among others.

There was specific interest in a strong focus on employee productivity in view of the transformation process under the new management. The rapid expansion of the business, involving takeover of towns' non-financially viable operations (as envisaged in the Strategic Direction), had an inherent requirement for employees who were committed and ready for rapidly emerging operational and managerial challenges. Moreover, the management arrangements for the new towns were bound to be characterised by low economies of scale and scope, hence focusing on wider corporate employee productivity would help to deal with this problem. In addition, subsequent employee remuneration designs had to take into consideration productivity considerations to assist in containing incremental operational costs. As a rule of thumb, the Strategic Direction envisioned the need to keep employee productivity at an average level of at most six employees per 1000 connections in the wake of rapid takeover of towns. This was to ensure that there was no performance retrogression (compared to past trends) while also remaining within remits of best regional practice. The remediation strategy was to encourage all regional operational managers to aggressively pursue business growth to buttress core staffing requirements.

Employee productivity management was pursued in a number of ways. All management and general staff meetings were restructured to include regular illumination of the vision and strategic direction performance expectations (goals and targets). The Managing Director took the lead in inspiring everyone to focus on the vision at every occasion of meeting staff and senior managers. In this regard, all weekly executive meetings had an agenda item for the Managing Director to give strategic guidance and inspire senior managers to pursue the corporate vision. At the same time, achievements and successes were highlighted and strategic risks pinpointed for remediation and management. Employees were continuously urged to put an emphasis on business growth as a

Figure 3.1 Dr Eng Silver Mugisha (left) hands over the trophy to the Manager Sewerage Services department right) in recognition for the outstanding performance of the department during the annual performance evaluation workshop at NWSC-IREC Kampala. August 2016.

sustainable way to create enhanced employee remuneration and career growth opportunities. Employee incentive plans were restructured to reflect new performance priorities and recipes for catalysing staff commitment. A performance-based salary scale structure (PSSS) was introduced in all operational regions, areas and branches. The PSSS had its rationale rooted in the need to remunerate managers and staff according to performance. This was carried out, taking into account achieved business size and geographical spread of operational businesses under their jurisdiction. The thresholds were designed to encourage vigilance and momentum to boost operational business growth. In addition, to forestall complacency and encourage thinking outside the box, managers and staff were continuously sensitized to employee productivity comparisons between areas/branches. Without this benchmarking approach, some managers were bound to think that their productivity levels were on the best-practice frontier, when it was actually far below peer threshold. Exhibit #17 highlights examples of efforts made by management to enhance staff productivity.

EXHIBIT #17: FROM MOSES BATEGEKA (2018), NWSC SENIOR MANAGER

Staff productivity has been enhanced through strong leadership and teamwork. The leadership offered has been so vital to the success recorded in the past four years. It has been very focused on achieving the organisation's mission (supplying water and ensuring supply reliability). Most staff were able to understand and articulate the strategic direction 2013–2018. The leadership has been very inspirational in guiding the organisation workforce and motivating the staff through many ways. Salary scale realignment to bridge the gaps in salaries for different scales which was conducted in 2013 was a significant game changer. This gave trust to staff that the new management was going to practice fairness and openness. Salary adjustment for all staff depending on level of education and length of service. This gave hope to staff that they will grow in the organisation. Structure reorganisation; this allowed vertical growth and created new positions which were filled internally, thereby motivating staff further. This was along with staff promotions to the vacant positions created and available. Hence the staff turnover reduced greatly. Staff development activities have taken place through fair treatment.

3.3 STAFF ENGAGEMENT PRACTICES AND WELFARE

Enhancing the concept of employee engagement was strongly envisioned in the new management's Strategic Direction. The intention was to revamp employees' orientation to work with passion and a feeling of profound connection to the organisation. Given the objective of rapid geographical expansion, the drive for innovation and motivation was paramount. Consequently, the new management had to devise means of enhancing the employees' emotional commitment the corporation and its strategic goals. This type of emotional commitment meant engaged employees actually cared about their work and their organisation. Exhibits #18–20 depict some insights into staff engagement practices since 2013.

EXHIBIT #18: FROM IBRAHIM WALUSIMBI (2018), NWSC AREA MANAGER

I have personally observed a number of activities to enhance staff engagement. These include: deliberate efforts to build the capacity and empower staff at the different levels in the corporation. In addition, entrusting leadership and managerial positions with our own internal talent, national vocational institutions in Gaba and Katchung, among other examples, speak volumes of this aspect. Visionary, solutions-oriented and fast-thinking leadership by the Board and, particularly, a vibrant, bold and results-oriented Managing Director who has been the pillar of transforming the corporation over the years has also played a key role in ensuring that employees get a strong sense of organisational belonging.

EXHIBIT #19: FROM STEPHEN WALAKIRA (2018), NWSC AREA MANAGER

In the previous four years, information flow concerns and gaps have been addressed to satisfaction. This has made the staff appreciate how changes would affect them personally. This developed staff confidence, hence enhancing quick decision making as they knew their expectations at individual and organizational level. Staff with personal concerns want to know how the change will play out for them. They wonder if they have the skills and resources to implement the change. It's normal for people to focus on what they are going to lose before they consider what they might gain. These staff concerns have been surfaced and addressed systematically by our leadership. This has in turn resulted into increased staff engagement and motivation.

EXHIBIT #20: FROM SHARON KARUNGI (2018), NWSC BRANCH MANAGER

Many groundbreaking performance improvement measures have taken place in the last four years. Prioritisation of staff welfare: issues regarding staff benefits and emoluments have been promptly handled. Improved salaries through scale alignment, promotions have empowered staff for better performance. Skills development: many more staff have undergone training and exposure (locally and internationally) in the last 4 years, including lower tier staff. Brand exposure: the NWSC brand is now a force to reckon with owing to deliberate efforts to boost the brand through social media. This has boosted customer confidence in the organisation.

The goal of these initiatives was to ensure that employees were up to speed with the corporation's vision, mission and performance objectives. We wanted to be sure that employees' interests and efforts were kept effectively threaded together. Unlike in the past, the Strategic Direction put emphasis on staff engagement as a key performance improvement strategy. The underlying conceptual framework for staff engagement was characterised by a number of building blocks. These included: alignment of the corporation to a common purpose of 'water for all', focus on collaboration, working together and putting an emphasis on corporate values. The values in the Strategic Direction included: result-orientation, integrity, excellence, professionalism, reliability, innovation and teamwork. Other practical building blocks included a variety of health and wellness activities (conducted regularly at the International Resources Centre and other venues), participating in corporate leagues, encouraging staff to undertake personal development projects. In the latter case, motivational speakers would be brought in to catalyse personal growth.

Other practices have included preparing and organising themed gatherings and specific days, such as breakfast meetings on every month-end Friday. Other such days include: end of year parties and performance evaluation workshops. On such days, workers that have shown exceptional performance and achievement are praised in public. In addition, corporate award recognitions and wins are announced and appreciated. The role of work–life balance is also underlined and employees are encouraged to practice it for effective productivity. Those days are also used as occasions to give workers 'inside' information in order to minimise knowledge asymmetry. For new employees, training/induction on corporation business is among the first conducted, including assignment of mentors and/or buddies for the continuous orientation of newcomers. Other staff engagement practices have included sending regular recorded e-messages to all staff from the Managing Director, with an emphasis on recent happenings in the corporation and planned activities/milestones. Senior managers are also required to give e-feedback (using a number of available digital platforms) in order to continuously motivate field managers and staff.

The other noteworthy aspect that has been introduced for staff engagement is a ***Revolving Fund*** facility. The Fund operates on the basis of a 5% (of employee gross salary) contribution by an employee with a matching 5% contribution by the corporation. In other words, the facility has an implicit 5% salary increment for staff. The contribution by the corporation is only transferred to employee ownership after some time as specified by the Board of Trustees of the Fund. The Fund is managed by an independent administrator sourced through a transparent competitive process. It also has a Board of Trustees selected from representatives of management and staff. The main objective of establishing the Fund is to enhance staff engagement, retention and welfare.

Furthermore, on the side of staff welfare, the new management, working with staff union management, negotiated a salary rationalisation and enhancement package. This step was taken on the basis that staff had stayed for a long time on almost fixed salaries, despite the performance efficiency gains that had been realised. The salary compression ratio – of the highest-paid (apart from the Managing Director) to the lowest-paid employee – was about 30:1. The idea was to reduce this disparity to about 15:1. Consequently, the lower-paid staff salaries were adjusted upwards by 50–80% in order to achieve the targeted compression ratio. The main objective of this rationalisation exercise was to improve employee buy-in, equity and fairness in the salary scale structure. The ultimate aim was to enhance employee engagement and motivation to embrace new management reforms by creating a cognitive recipe that when efficiency improves, both the corporation and employees win. Of course this policy was implemented at an initial cost of cutting back in the corporation surpluses, as an investment in human resource capital, to build the necessary start-up momentum.

3.4 NATIONAL, REGIONAL AND INTERNATIONAL RECOGNITION

Owing to the good performance during 2013–2018, NWSC has been recognized at the national, regional and international level. Table 3.1

Table 3.1 Awards obtained by NWSC during the period 2014–2018.

Award Description	**Year of award**
Best Government Procurement Complaint Entity awarded by PPDA of Uganda	2014
Manager of the Year Award to Dr Silver Mugisha by European Quality, Oxford, UK	2015
Best Financial Reporting in the Government category awarded to NWSC by Institute of Certified Public Accountants of Uganda, in collaboration with Uganda Capital Markets Authority and Uganda Securities Exchange and	2015 and 2016
World Leader Business Person awarded to Dr Christopher Ebal and Dr Silver Mugisha	2015
Best Customer Care Entity awarded to NWSC, organised by Uganda Social Media Awards	2015 and 2016
Employer of the Year Award in Learning and Development awarded to NWSC, organised by Uganda Federation of Employers Union (UFEU)	2015
Africa Water Utility of the Year Award to NWSC, organised by Africa Utility Week, Cape Town, South Africa	2014, 2015, 2016
Gold European Award to NWSC for Quality and Commercial Excellence, organised by Otherways Management and Consulting Association of France	2015, 2016
ERA Gold Award to NWSC, organised by Century International Quality ERA	2015
International Diamond Excellence Award to NWSC, organised by European Society for Quality Research	2015
Peak of Success – the BIZZ 2016 and Excellent Service Award to NWSC, organised by World Confederation of Business (WORLDCOB)	2017

(*Continued*)

Table 3.1 Awards obtained by NWSC during the period 2014–2018 (*Continued*).

Award Description	Year of award
Best Digital Customer Service by Utilities Brand Award to NWSC, organised by HiPipo, Uganda	2017
Best Financial Reporting in the Government category Awarded to NWSC by Institute of Certified Public Accountants of Uganda, in collaboration with Uganda Capital Markets Authority and Uganda Securities Exchange	2017
People's Choice Award to NWSC, organised by Uganda Public Relations Association	2018

shows some of the awards that the corporation has obtained during 2014–2018.

The above awards have helped in a number of ways. First, they have helped to galvanise the spirit of success among staff. Management did

Figure 3.2 Dr Eng Silver Mugisha receives the Africa Leadership Award 2016 for his exemplary leadership at hotel Serena Kampala. June 2016.

Figure 3.3 NWSC board and management receive the African Utility of the Year 2014–2015 Award in Cape Town South Africa. May 2015.

Figure 3.4 NWSC board chairman – Dr Eng Christopher Ebal (third from left) and corporation secretary – Ms Edith Kateete (fourth from right) receive the Bizz Award, awarded to NWSC for excellency in service delivery in Monaco France. May 2016 on behalf of the Corporation.

not put a lot of emphasis on how the awards were being given to NWSC, but more on the message and motivation they brought with them. We did not want to suffer the common disease of complacency that comes with such awards but rather foster a spirit of propensity to achieve yet more. The second attribute that came with these awards was a perception of authenticity among external stakeholders that the management reforms being undertaken were effective and valid. The third attribute was the message of change management and continuous improvement that accompanied these awards. It was much easier to gain staff buy-in to proposed restructuring: after all, even outside organisations put a 'stamp' in them. In other words, management used these awards as external proof of its effective agendas. The fourth attribute was a message of consistency: the fact that awards have consistently been coming to NWSC, in respect of different aspects of management reforms, has shown a perceived sense of consistency, continuity and effectiveness. Exhibit #21 presents some recipes to winning various awards by the corporation.

EXHIBIT #21: FROM STEPHEN WALAKIRA (2018), NWSC AREA MANAGER

I attribute continuous winning of awards to a number of ways of creating sustainable success, innovation and creativity. Once performance initiatives were initiated, NWSC leadership were open minded to listen to others about how attendant changes can be sustained. For example: financial viability, meeting political expectations while balancing with improved service delivery and making the change even better. Leadership discussed issues to do with staff motivation, staff productivity, involvement, supporting innovativeness and creativity in the different departments of the corporation. During the course of NWSC change management initiatives, a number of lessons have occurred. Many staff took advantage of new opportunities that were created as a result of business growth. We have harnessed cultural diversity and regional balancing that has created a sense of ownership among staff, communities and different stakeholders.

Figure 3.5 NWSC Corporation Secretary – Ms Edith Kateete (Left) and digital media team receive the Best use of social media award by a government parastatal during the social media awards in Kampala. April 2016.

3.5 INCULCATING A CONTINUOUS LEARNING CULTURE

The mantra of continuous learning was envisaged in the Strategic Direction. This aimed at strengthening abilities to progressively develop and improve employee skills and knowledge for purposes of effective performance and adaptation to changes in the work place. Continuous learning was framed on two levels: the employee level (both personal and professional) and the corporate level. On a personal level, the new management was concerned with the employees' constant expansion of skills/practices and skill-sets through learning. In this regard, employees were encouraged to pursue a multitude of career growth agendas. These included incentivising staff to go for further studies, refresher courses and develop a self-driven culture of always looking for new knowledge through reading. Incentives included: support in acquisition of study loans, flexible leave periods and working times, research grants and discretionary field allowances. Reading/reference materials were made more readily available through a strengthened library, internet facilities and subscription to peer-reviewed journals and scholarly magazines. On the innovation side, the human resource manual was revised to include incentives for employees who engage in creation of value-adding innovation products.

On the professional level, continuous learning was about further broadening of skill-sets in response to changing operating environments and emerging challenges. In this case, employees were encouraged and supported to undertake professional courses leading to membership of professional bodies like the Association of Chartered Certified Accountants (ACCA), Chartered Public Accountants (CPA), Uganda Institute of Professional Engineers (UIPE), Engineers Registration Board (ERB), Institute of Company Secretaries (ICS), among others. Management provided incentives to encourage staff to undertake professional courses by paying subscription fees, making available study loans and giving moral support. Staff knew that acquisition of such qualifications would accelerate their chances of promotion. Within the period 2014–2018, the number of employees with professional qualifications more than doubled. This has created reciprocal positive effects on the performance of the corporation as exemplified by the number of awards obtained and the speed of work in response to ever increasing workload from a continuum of managerial reforms.

At the corporate level, there are many facets of continuous learning that have been introduced by the new management. A plethora of performance improvement programmes now exist in different areas of operation in response to the ever increasing and changing performance challenges and risks. This has progressively created a fertile ground for learning lessons and revealing new ways of addressing emerging issues. The main strategic focus areas of the performance programmes have included: customer service improvement, water loss prevention, revenue improvement, community mobilisation and connectivity, cost optimisation and corporate governance. As a result of the design and implementation of different performance improvement programmes, the corporation and its employees have adopted a strong culture of change management and learnt how to live with the reality of change as the only constant factor. Moreover, relentless efforts by the

Managing Director and senior managers to send a message of the need to always challenge the status quo, through meetings with staff has helped to boost this culture of continuous learning and improvement. Having continuous improvement as part of the main theme of the Strategic Direction has also helped to reinforce a culture of continuous learning.

The establishment and operationalisation of a Vocational Skills Institute (VSI) at Ggaba Water Works, specifically to deal with imparting skill-sets to the technician segment of staff has also helped to enhance a culture of continuous learning. The VSI is run in partnership with Uganda Business Technical Vocational Education and Training (BTVET) and offers accredited courses to NWSC staff. The courses are diverse in nature including: plumbing, electro-mechanical technology, customer care, marketing and meter reading. A bigger vocational skills training facility is being built in Lira (one of the operational regions of NWSC, located in Northern Uganda) to cater for expanded training needs in Uganda and the region. The VSI and its attendant operations were established in 2013, with support from German Development Co-operation through GiZ of Uganda.

Apart from the VSI, the corporation also operates an International Resources Centre (IREC) that offers a variety of short courses, mainly for middle and senior managers. The courses, which are powered by an active Research and Development (R&D) department, are offered partly in collaboration with international training institutions. These include: IHE-Delft, Netherlands; VITENS-Evides International, Netherlands; and Loughborough University in the UK. The corporation, using its internally generated resources, has recently started sponsoring targeted employees for further studies in international universities in specific courses required for its effective operations. This is done for middle managers with unique potential abilities who have demonstrated good performance: those who are promising star performers. The selection is augmented by a competitive process and regionally balanced. Successful candidates have to accept a bonding arrangement before being considered for sponsorship. This is because the corporation wants to derive maximum return on study investments undertaken in terms of service to the corporation after completing studies. Exhibit #22 gives more insights regarding inculcating a continuous learning culture in the corporation.

EXHIBIT #22: FROM RONALD ODHIAMBO (2018), NWSC BRANCH MANAGER

In my view, a continuous learning culture and propensity to win awards has been precipitated by a number of initiatives in the last four years. The timely execution and completion of projects using internal funds e.g., Naguru and Kanyanya water projects to improve service delivery. This has created a culture of in-house design and construction without recourse to external partners. We have built self-awareness through increased stakeholder engagement and feedback thru meetings guided by the Managing Director, enabling timely feedback from the stakeholders to improve service delivery. These include: barazas and meetings with local leaders to share experiences and communicate NWSC plans for improvement. In addition, a culture of probing how customer and the public feels about our offering has been created. This has been through adoption of social media for both external and internal communication, where real time sharing of on-going events outside the country and within. We also see opportunities given to staff who are not part of top management to travel for international workshops and events to enable learning through benchmarking, which was previously limited to a few.

Figure 3.6 NWSC Staff pose for a photo during the first graduation ceremony at the Ggaba Vocational Skills Institute Kampala. September 2017.

Figure 3.7 The Minister of State for Higher Education (third from right) together with NWSC board and Management presiding over the graduation ceremony of NWSC staff of the Ggaba VSI Kampala. September 2017.

Figure 3.8 NWSC Manager Training and Capacity Development – Dr Martin Kalibala (first right) conducting refresher training for NWSC commercial staff at IREC Kampala. June 2017.

3.6 OTHER STAFF MOTIVATION TECHNIQUES

As pointed out above, the spirit of the Strategic Direction was a complete rethink agenda seeking to create stepped performance achievements. The message of the Managing Director to all managers and employees was to work as if 'there is a lion in the compound'. This called for a strong sense of urgency in action, guided by discipline and commitment. However, as expected, as leaders, we could clearly see that this performance spirit required us to create strong inspiration to achieve the collective vision by talking about it every time we got a chance to address staff. We had to use the approach of telling staff that our survival in the eyes of politicians and other policy makers was to aggressively target those without water and sewerage services while consolidating gains realised already. The message of the Managing Director was 'don't be like a mad man' who does the same things everyday expecting different results. Hence, there was need to continuously challenge the status quo. We were thirsty for new remedial plans to respond to ever changing performance challenges. Using the newly launched Performance, Autonomy, Creativity and Empowerment (PACE) contracts, we put particular emphasis on empowerment of managers and staff and encouraged them to work as if there were no managers at headquarters. Some of the activities that were delegated to managers to carry out at regional level included: land title acquisition, handling legal cases, environmental protection, human resources management and capital investments. Managers were encouraged to stop complaining about what was not going on well but 'take the bull by its horns' and make things happen on their own. The Managing Director, specifically, emphasised that leadership is an action and not a position and that managers were to be evaluated based on their practical actions not mere talking. Public relations management based on 'empty' talk was discouraged and replaced by action-based public relations management. Managers were continuously urged to lead by example in terms of time keeping, attention to detail, honesty, mentoring and coaching staff and dress-code. Exhibit #23 shows how actions can

EXHIBIT #23: FROM GRACE LWANGA (2018), NWSC BRANCH MANAGER

In the last four years, I have personally, been motivated by the notable achievements I have observed. These include: rapid growth in towns served that is from 23 to over 225 (and still counting); creation of SWAS clubs that have been of great importance in nurturing the young generation; introduction of highly innovative water for all project code-named SCAP100 where over 7000 km of water mains are to be constructed and 12,000 village cells served; introduction of a vocational skills training facility in Ggaba to equip technicians (mainly plumbers) with adequate knowledge for better service delivery; the corporation receiving a number of recognition awards (locally, regionally and internationally; our Managing Director being elected Vice President of IWA.

also motivate employees and increase their sense of belonging to the institution.

The other strong motivating factor was putting an emphasis on effective communication and persuasion. The new management was aware that an employee can have ability but no willingness to perform. A strong corporate-wide persuasive framework, therefore, had to be developed and used to cascade a plethora of performance expectations to staff. There were a number of key ingredients in this framework. We told staff that if the corporation achieves, they will share in terms of increased incentives and career growth opportunities. On the other hand, we pointed out that if performance went down, government may resurrect the idea of privatising the corporation and undertake staff redundancy reforms. Clearly, the issue of reciprocity and scarcity of jobs were key motivators in this regard. At that time, the Managing Director had a published a book titled *Utility Benchmarking and Regulation in Developing Countries: Practical Application of Performance Monitoring and Incentives*. This book was distributed to staff at subsidized rates as a credible expert tool to catalyse action based on past experience in the region.

As senior managers, we also identified a number of examples demonstrating how consistent commitment to managerial reforms had yielded good results in the past. Some notable examples were the stretch and one minute programmes that had been successfully implemented about 10 years ago. Consequently, another programme called Sky Performance Programme (SPP), based on stretch principles of speed, self-confidence, simplicity and stretch goals was designed and implemented in all areas in 2017. The programme aimed at unpacking and restructuring motivational software issues amongst employees. In addition, we also used key champions in the organisation, with likable character and in some cases similar cultural background, to bring on board disengaged employees to embrace changes. Sometimes, in-office guest motivation speakers were also called in to discuss previous success stories akin to managerial changes proposed with an aim of building consensus on reforms. This was done, specifically, in situations where we had 'doubting thomases', who could not believe that some ambitious reforms (like rapid take-over of unviable towns, tariff rebalancing etc.) were possible and had ended up in a disengaged disposition. Exhibit #24 depicts how explaining organisational changes to staff can lead to their effectiveness.

EXHIBIT #24: FROM STEPHEN WALAKIRA (2018), NWSC AREA MANAGER

The current leadership addressed the first two concerns effectively, staff were prepared and ready to hear information on the details involved in implementing the change. At this stage staff were interested to hear how the thinking behind the change could be tested. With restructuring and clear policies, guidelines put in place, staff know where to go for technical assistance and solutions to challenges as they might arise. Because of the improved efficiency in information dissemination, feedback can easily be relayed to staff queries even to bottom staff by introducing programs like Stretch-out and Sky Performance Programme that address their concerns. Clear targets were set and steps laid how to achieve them.

3.7 A NEW APPROACH TO STAFF APPRAISAL

At NWSC, we use staff performance appraisal as a periodic process of assessing individual employee's job performance and productivity in relation to certain sets of criteria and organisational objectives. Prior to 2013, employee appraisal was based on a two-way system: the employee appraising themselves against pre-set performance targets and harmonising this with the employer. The main challenge with this system was two-fold. First, the targets were largely lacking in measurability and clarity. This, most often, bred subjectivity and resentment on the side of the employee being assessed. Second, a significant part of the appraisal system was heavily based on individual performance and was poor at assessing employee potential in terms of competences (knowledge, skills and attitude). The modern way is to assess both perspectives (performance and potential/competence) so that the appraisal system can help to identify which employees are 'deadwood', 'solid citizens', 'question marks' or 'stars'. This identification process, if properly carried out, can help to determine the level of stratification of investment in human resources development for effective productivity management and succession planning.

After introducing the Strategic Direction, we modified the employee performance appraisal system to incorporate evaluation of the two perspectives: performance and potential. Performance appraisal was based on prior-formulated Individual Performance Agreements (IPAs) with clear individual performance targets and timelines for achievement. The competence-based appraisal was carried out through a 360-degrees appraisal system where an employee would be appraised using information from the superior, line and junior staff (top-down-horizontal approach). During this form of appraisal, an employee would be presented with competence-based information on issues of honesty, coaching ability, forward-looking orientation, intelligence, capacity to inspire others, social skills, and empathy, among others. The appraisal information was partitioned into relative strengths and weaknesses. The employee would then be asked to respond to an appraisal team panel. Some of the information would be perception and some would be facts, but the employee would be asked to take a positive approach, bearing in mind that perceptions can also represent reality. Our findings throughout these appraisal sessions show that employees are happy to be told who they are and how they can improve. They have often credited this approach as transparent, fair and forward looking in terms of helping the employee's personal development efforts.

With the two fronts/approaches to performance appraisal, the corporation is now better placed to know who are the key employees that form the fulcrum for sustainable performance improvement. This has also helped to instil a strong culture of career growth orientation among employees, aligning them better to key competences that are necessary for successful personal development. There has been a strong shift from over-emphasis on knowledge (formal education) to incorporating a focus on skills development and attitudinal change as well. Previously, employees were liable to think that when they acquires additional educational qualifications, they were entitled to automatic promotion. We have now instilled a culture of appreciating that it takes more than just academic qualifications. Actually, some schools of thought posit that behavioural attitude contributes more than 70% of what is required for an employee to succeed in organisational work endeavours. Indeed, there are corporate plans to re-orient training programmes to incorporate all these human resources development dimensions. Exhibit #25 highlights some measures taken to enhance staff performance accountability.

EXHIBIT #25: FROM MOSES BATEGEKA (2018), NWSC SENIOR MANAGER

In the last four years, we have seen both internal and external accountability. Internally, there has been several and consistent staff appraisals at various levels conducted. Individual performance agreements were made and periodically assessed. These reminded the staff of their obligations. This system of assessment enabled feedback to staff and kept them aligned to their commitments. Targets have been consistently formulated and performance evaluations held periodically. This has created competition and increased organisational performance. Externally, more than ever before, the organisation has been able to listen to the customers through the WACOCO programs (messages are spread to the customers at market places, churches and other forums) and barazas. This has been aided by the social media platforms; the ability to change with the wave of technological changes enhanced success of the organisation. Nowadays, you cannot talk about what you do not have and promise what you will not deliver in service industry. Staff and managers are also evaluated on their effective handling of external customers.

Chapter 4

Financial perspective

This perspective of the Strategic Direction put specific emphasis on financial performance: the processes, systems and other issues related to the stewardship of the corporation. Management had the duty to consider key value-adding fundamentals aimed at yielding financial outcomes that would please government as the shareholder. This section presents, in brief, those key measures, attributes and facets that were incorporated in the Strategic Direction and subsequently implemented. Exhibit #26 gives examples of recipes for improved financial performance of NWSC.

EXHIBIT #26: FROM CHRISTINE AWINO (2018), NWSC BRANCH MANAGER

Over the last four years, the financial muscle of the corporation has been strengthened through various operational and process activities. These have included implementation of infrastructure service delivery plans and water supply stabilisation programmes – this has almost eliminated dry zones in Kampala (e.g.,) and hence

Author: Silver Mugisha
doi: 10.2166/9781789060256_0089

increased water sales; the sewerage network has been expanded and hence increased revenue from sale of sewerage services; improved customer engagement thereby creating more willingness to pay and hence enhanced collection efficiency; in-house capacity building programmes thereby creating better financial discipline and understanding.

4.1 EMPHASISING SERVICE EXPANSION RATHER THAN RETAINED EARNINGS

When NWSC was only responsible for Kampala (the capital city) and a few other towns, managers focused on retained earnings which could ultimately be ploughed back into the areas – addressing capacity-building, network maintenance, and sanitation. The key distinguishing feature of the new Strategic Direction compared with previous strategic plans was the overall objective. That objective of service expansion presented new challenges for managers. However, since the organization had successfully moved from a relatively small, high-cost utility, it was in position to emphasize network expansion in less densely populated towns that were not breaking even. It was ready to move beyond improving cash flows to applying efficiency gains to financing capital outlays and maintenance expenditures. The savings had partly been achieved through scale economies in larger (and more densely populated) towns. Now they were applied to towns and villages that had not been part of NWSC in the past. These smaller utilities lacked the managerial and technical skills necessary to increase their cash flows. At the same time, their placement under direct jurisdiction of local political structure posed governance challenges.

By cutting back on overall potential corporate profitability, NWSC was able to rapidly increase the number of towns under its organizational umbrella; NWSC was in a position to serve more people and provide better service. Better service means better customer care, increased water supply reliability, more dependable functionality within water production systems, better financial and billing systems, increased network expansion (increased new connections in the towns), increased focus on the poor and vulnerable citizens, greater attention to stakeholder engagement and more avenues for promoting accountability.

Figure 4.1 Dr Eng Silver Mugisha (left) hands over a bucket of water to the resident Kanungu during the launch of public water point in Kanungu area as part of the service expansion programme. March 2017.

In addition to operational expansion, the Strategic Direction established an aggressive investment program to expand networks, even in towns with predominantly low-income communities (what profit-oriented service providers would call 'social-mission' programmes). The initial result of such development programmes would be low levels of returns on investments. However, the strategy was designed to improve cash flows later when connection hook-ups and water consumption expanded. In such cases, the main mode of service was Public Stand Posts (PSPs) that served 20–30 homesteads each (about 200–300 people), obtaining water within a maximum distance of half a kilometre.

4.2 OPERATING COST CONTAINMENT MEASURES

Since 2013, with the promulgation of the Strategic Direction, the Board and Management have emphasised network expansion (investment) rather than recurrent expenditures (variable costs). Annual budgets are implemented in a flexible manner, cutting back on consumption (wasteful) expenditure to apply cash flows for network expansion and water supply stabilisation programmes. This approach is implemented carefully, mainly targeting reduction of administrative expenses, while emphasising operation and maintenance of core infrastructure. Staff expenditure, inland/foreign travel, meetings and similar expenditures have been drastically reduced to increase funds for capital expenditures. Consultancy services, security expenses and billing system licenses have been contained. Accordingly, the corporation's IT software applications development team developed a new customer billing system, based on the web (instead of relying on a license-based system). Of course, funds for recurrent expenditures remain important: maintenance and other activities have implications for future financial sustainability. However, after much discussion, managers determined that the sources of such funds should be from continuing improvements in cost containment, reduced non-revenue water, and increased collections.

A number of initiatives have been put in place to leverage more financing to make the corporation more self-reliant. These have included a number of initiatives: significant energy reduction plans in water treatment plants through innovative engineering; introduction of e-payment systems using banks and mobile money; e-procurement systems that eliminate excessive paperwork; a home-grown web-based billing system that eliminates payment of license fees; and performance-based manpower planning arrangements. As pointed out earlier, a plethora of water loss prevention units (WALOPUs), with varying forms of incentive plans, were established in all NWSC utilities for this purpose. This approach involves working with police, other security agencies and local communities to combat the illegal connections/water thefts that are the major sources of commercial losses. In addition, NWSC introduced on-spot billing systems using smart phones to reduce the cost of customer invoicing and attendant errors.

4.3 REDUCING UNIT INVESTMENT COSTS

Unit costs of capital investment are always an issue of concern in most water companies in developing countries. As of June 2013, for example, NWSC's average capital investment unit costs for internally funded projects were in the range of Ush 800 (US$ 0.30) per unit of water sold. The main cost drivers were high consultancy costs (despite availability of internal design capacity who were willing to do the job), high compensation costs, exaggerated labour costs, high costs of materials due to unplanned procurements, high overheads and contingencies due to collusion between managers and contractors, among others. The new management had to deal with these issues in an innovative way.

There were a number of remediation measures implemented to bring down unit investment costs. Engineers were incentivised (through a number of emotional incentives such as recognition, promotion and trophies, among others) to carry out field work, feasibility studies and detailed designs internally. This helped to cut down on outsourced consultancy fees. In addition, through better collaboration with local political leaders and communities, the corporation was able to implement projects with compensating beneficiary communities. Labour costs quoted by contractors have also been reduced through use of direct labour under a 'force account' mechanism, a procurement method that is provided for under the law. Furthermore, materials costs have been reduced through extensive use of framework contracts with pipe factories and other materials suppliers. This system allows quick re-order processes using call-off orders. Collusion with contractors has been greatly curtailed through use of direct labour contracting and vigilance by internal audit department, which has since been strengthened through introduction of periodic value for money audits. As a result, as of June 2017, unit investment costs had reduced to about Ush 600 (US$ 0.16) per unit of water sold. This has helped to leverage more investment financing for increased infrastructure expansion. The efficiency gains from improved unit investment costs partly explain why the rate of new pipeline extension has increased from about 120 km per annum in 2013 to more than 1000 km per annum in 2017. Exhibit #27 illustrates how changes in procurement methods have increased project implementation efficiency.

EXHIBIT #27: FROM MOSES BATEGEKA (2018), NWSC SENIOR MANAGER

The organisation has had a very good financial capacity within the last four years (we can call it financial discipline) than before. Many projects aimed at water stabilisation and extensions (increasing water coverage), were implemented, particularly in Kampala City. Dry zones have become history out of undertaking these projects and water services were extended to unserved people. Most of these projects have been undertaken using internally generated funds.

> Operational flexibility has reduced bureaucracy. Enforcement of direct-labour-based method for project implementation (projects undertaken using the force account procurement method) increased speed of project delivery. Previously, we followed the bureaucratic outsourcing to contractors that was so laborious and ineffective. This blueprint approach which has had to be meticulously followed led to many implementation challenges. There were delays in execution of projects due to a lengthy period of obtaining contractors – usually it would take more than three months and in some cases the process could overlap financial years. We also had very costly projects both in terms of time for execution and cost of the actual project. Moreover, there was mishandling of information about projects by contractors, causing inaccurate project documentation. The direct labour method closed the above gaps identified in the traditional methods of labour works contracting, increased speed for projects implementation and enhanced staff skills.

4.4 RATIONALIZING THE TARIFF STRUCTURE

The current NWSC tariff structure came into force in April 1994. In setting the tariffs, the corporation is guided by a number of principles. Given that not all the NWSC towns are able to break even, that is cover operation and maintenance costs, the corporation operates a uniform tariff structure across all its towns. Water is essential to life (and so can be viewed as a basic right) and hence consideration must be taken to allow subsidization of minimal water needs to poorer customers. Thus, tariffs need to take into account the customers' ability to pay. In addition, for financial sustainability, tariffs are set such that revenues generated cover operation and maintenance costs, replacement of assets, debt servicing and expansion. Furthermore, tariff setting takes into account the fact that the corporation has the responsibility to provide free water for humanitarian purposes to society in particular cases, for example for fire hydrants.

4.4.1 Tariff setting procedure

The setting of water tariffs is governed by the Water Act Cap 152, Section 94 and involves a number of steps. The corporation management and Board prepare a submission to the Minister of Water and Environment indicating and justifying the suggested tariff revision. The Minister evaluates the proposal (with the technical assistance of the Regulation Department in the Ministry of Water and Environment) and amends or approves it. Afterwards, a Statutory Instrument effecting revision of the tariff is issued. In essence, therefore, the tariff is regulated by the Ministry of Water and Environment, acting as the Economic Regulator.

4.4.2 Tariff policy in a historic perspective

From 1994 to 2000, the tariff remained unchanged and was therefore eroded over time by factors such as inflation and depreciation of the shilling against the dollar. In 2000, it was estimated that the tariff value was only 45% of the real value in 1994. In 1998, the NWSC management engaged the services of a consultant to review the tariff structure. The objective of this review was to improve the revenue base of the Corporation by raising customers' willingness to pay. The study recommended, among other things, an 18% increase in the tariff if the NWSC was to effectively cover its operational costs. This was not done, however, owing to the intervention of Parliament. A few recommendations of the study were subsequently adopted in the year 2000. These included the removal of minimum charges, the reduction of the connection and reconnection fees, and the introduction of a service charge.

In April 2002 Cabinet passed a policy of tariff indexation to protect the tariff from loss of value due to changes in domestic inflation, electricity tariff (energy charges), foreign inflation and exchange rate changes. The indexation policy aimed to enable NWSC to cover its operational costs, thereby ensuring sustainability of service provision. It should be noted that despite the indexation policy, the NWSC tariff was still not to a full cost recovery level. This means that it could cover operation, maintenance and depreciation costs, but was not able to cover full financial costs (debt servicing) and return on investments. In addition, in 2004 and 2006 respectively, the NWSC introduced simplified water and sewerage connection policies in which new water and sewerage customer connections were subsidized within a radius of 50 and 60 m respectively. This move improved access to the NWSC services and at the same time reduced incidences of water leaks and bursts on consumer service pipes up to the consumer water location. With effect from July 2015, the corporation introduced an infrastructure fund to take care of short–medium capital investment requirements and also build a financial base that could ultimately be used to match funds from national and international capital markets. The contribution to the infrastructure fund stems from a tariff adjustment where a new connection policy was introduced, requiring the wealthier customer segments to pay for their own connections, leaving the corporation to provide only free public stand posts. The other contribution is from 10 percent tariff adjustment, implemented over two years. All infrastructure fund resources are ring-fenced for capital development activities.

In its quest to further improve accessibility, as well as implement the Presidential Directive of ensuring universal access to clean water by all cities, towns, trading centres and villages, the corporation has, since July 2017, been implementing the three-year project codenamed '100% Service Coverage Acceleration Project (SCAP100)'. The project is funded by NWSC (60% of the budget) together with the Government of Uganda (40% of the budget) and aims at 100% provision of safe water, targeting to cover at least 12,000 villages. Among other interventions this entails undertaking accelerated installation of public/social taps targeting villages/cells under NWSC's jurisdiction (2 public taps per village or a tap per 200 people). The corporation's capacity to finance SCAP100 was enhanced through a further tariff rebalancing. The main objective of this rebalancing was to improve affordability for low-income communities (tariff reduction of about 30%) and generate more revenues by adjusting the domestic tariff upwards (by about 20%). The rationale behind this was that domestic customers (relatively high-income category) were paying much less that the vendor rate for the low-income category owing to the scarcity of social connections.

4.4.3 The NWSC tariff structure for 2017–2018

The NWSC tariff structure (see Table 4.1) is skewed with the public tap (PSP) and domestic customers paying less than the institutional and commercial customers (except fpr commercial consumption exceeding 1500 m^3/month). As already pointed out above, the NWSC has a uniform tariff across all its towns. Scale economies, different customer densities, topology, and network age all affect costs. The tariff, therefore, has an implicit cross-subsidy across towns and within customer categories. The sewerage tariff is based on the volume of water consumed, (i.e., 75% for domestic and 100% for commercial).

Table 4.1 NWSC tariff structure in financial year 2017–2018.

Consumer Category	Tariff FY2017–2018 (Ush/m^3)	Tariff Per Jerrycan (Ush)	Tariff Per Jerrycan with VAT (Ush)
Public stand post	1060	21	25
Domestic	3305	66	78
Institutions/Government	3344	67	79
Commercial < 500 m^3/m	4102	82	97
Commercial 500–1500 m^3/m	4102	82	97
Commercial > 1500 m^3/m	3278	66	77
Average commercial	3827	77	90
Average tariff	**2855**	**57**	**67**

Note: 1US$ is equivalent to about Ush 3650 (as at February 2018).

4.4.4 Revenue from each customer category

Table 4.2 below shows the various market segments served by the NWSC. Statistics as of June 2017 show that domestic connections account for the highest percentage of the customer base (81%) and commensurately the highest volume of water billed 50% (providing 40% of the revenue). The institutions/government category account for 3% of the connections and 18% of the water billed (23% of revenue), while the commercial/industrial account for 14% of the connections and 35% of the revenue. Table 4.2 alludes to the fact that all customer categories and their current pricing are critical to the financial sustainability of the Corporation.

Table 4.2 Water market segments as of June 2017.

Consumer Category	No of Connections	As % of Total Connections	Volume of Water Billed (m^3)	As % of Total Billed Volume	Revenue Billings, VAT-Inclusive (Ush Million)	As % of Total Revenue
Public stand posts	10,424	2%	3,803,554	5%	7175	2%
Domestic	427,258	81%	40,102,727	50%	138,445	40%
Institutional & government	15,714	3%	14,719,647	18%	7875	23%
Industrial & commercial	71,261	14%	21,509,071	27%	122,435	35%
Total	**524,657**	**100%**	**80,134,999**	**100%**	**346,807**	**100%**

4.4.5 Basis for the tariff structure

The NWSC tariff is based on a number of tenets. First is the equity consideration, which is premised on the need to provide all sections of the population with at least the basic supply of water services. The poor are a key consideration in this respect. This objective is attained through differentiated tariffs for the different customers. Currently, the NWSC tariff caters for the equity principle with the price at the stand posts and domestic consumers paying less than the other categories of consumers. Furthermore, the tariff is an increasing block structure aimed at ensuring equity between the various users. The second criterion is based on the need to cover costs of operation referred to as the financial pricing criterion. In simple terms, the NWSC tariff is aimed at generating sufficient revenues to cover operation and maintenance (O&M) costs, depreciation and yield an acceptable return on assets.

4.4.6 Is NWSC tariff full cost recovery?

Currently, the NWSC tariff does not cover full costs (operating and capital costs) as it is only able to cover O&M, depreciation and minor investments. This is due to a number of factors including the increasing costs of production due to macroeconomic factors, and majorly due to the fact that the NWSC has to cross-subsidize many of the smaller towns which currently do not break even. Table 4.3 (a) and (b) below summarizes the relationship between the tariff bands and the average unit cost of water sold.

Table 4.3 (a) and (b) show that the current stand post/social connection price is highly subsidized, while the tariff for other consumer categories covers the O&M costs plus depreciation and minor capital expenditure costs. Because social connections are highly subsidized, the average tariff can only cover the operating costs, depreciation and a small fraction of

Table 4.3 (a) NWSC tariff structure vs unit cost of production.

Cost Implication	Tariff (Ush/m^3)	Unit Cost [Opex] (Ush/m^3)	Unit Cost [Opex + Depr.] (Ush/m^3)	Unit Cost [Opex + Depr. + Minor Capex] (Ush/m^3) (Ush/m^3)
Average Value	2855	2259	2538	3124

Note: 1 US Dollar is equivalent to about Ush 3650 (as at February 2018).

Table 4.3 (b) NWSC tariff structure February 2018.

	PSP	Domestic	Commercial	Industrial
Tariff (Ush/m^3)	1060	3305	4102	3344

minor capital investment costs. Therefore, the NWSC tariff structure is premised on the proposition of ensuring equity and balancing customers' ability to pay.

4.5 FINANCIAL AND SOCIAL ASPECTS IN INFORMAL AREAS

As already highlighted above, water and sanitation services without due regard to all categories of people do not provide a total solution for health improvement. NWSC takes the matter of service to all customer segments seriously. In this case, the income levels come into play. The tariff structure is differentiated to keep affordability of services. The water vendor effects, where prices are multiplied to five to ten times the official price of the corporation, are reduced through use of pre-paid meter technologies in slum areas of Kampala and through smart price caps for public water points in other operational towns. Continuous sensitisation of the population in low-income settlements, where education levels are low, also allows effective uptake of water and sanitation services. Effective coordination with city/urban authorities also ensures that operational and jurisdictional conflicts are efficiently identified and resolved.

4.6 KEEPING CLOSE WATCH ON REVENUE ACTIVITIES

Revenue generation is an important cash-flow improvement measure for any organisation. The successful implementation of the Strategic Direction at NWSC required specific priority emphasis on revenue management. There were a number of performance drivers that had to be emphasised to create a difference in revenue management. First, we had to make sure that the tariff structure was applied correctly among the different customer categories and billings monitored daily, via an email system, during the entire monthly cycle to avoid staff laxities. The idea of daily monitoring of the billings was a new one, meant to manage exceptions that could potentially impact negatively on total billing. In the same connection, bill accuracy was improved to enhance customer willingness to pay. This has been greatly improved through universal on-spot billing so that the customer gets a bill at the time of meter reading and has time to raise queries, if any. The introduction of instant on-line relaying of meter readings directly to the billing system using smart (rugged) phones has also improved billing efficiency.

On the revenue collection side, daily targets (derived from monthly targets) are set and monitored by a dedicated team in the Directorate of Commercial Services. Performance gaps are promptly identified and communicated upwards and downwards through specific email communications to field managers. This approach also allows cross-utility performance comparisons, which breeds a spirit of competition and benchmarking. Responsible senior managers also keep a close watch on these performance trends and offer guidance from time to time. This strategy helps to keep field teams on their toes, which has helped to yield consistently high collection efficiencies. In the same connection, revenue collection management has been structured through decentralised units/branches, which are linked to their regional offices and eventually to headquarters by a chain of performance contracts. Commercial officers in these branches are the main focal persons that coordinate revenue generation activities. The use of regional WhatsApp platforms and the requirement to post individual branch performance is also an important performance driver. Exhibit #28 further illustrates recipes for strengthening revenue generation activities of the corporation.

EXHIBIT #28: FROM MOSES KAIJUKA (2018), NWSC BRANCH MANAGER

I have made observations regarding the way the corporation staff have kept a solid grip on revenue and finance management, generally. We have adopted extensive use of social media platforms like Facebook, Twitter and Whatsapp (internally and externally) which has created personal as well as public accountability of corporation staff. This has led to instant customer engagement and feedback which has in turn improved effectiveness, efficiency in service delivery. We have also inculcated a strong nurturing and development of a 'can do it' attitude – staff have been empowered with an attitude to achieve what they set out to do as a must, without reservation. This positive attitude has cascaded from top to bottom in the corporation and this is attributed to the able leadership at the top. We have also developed a specific emphasis on internal financing as opposed to only concentrating on external and donor funding. This strong emphasis on self-financing orientation has enabled the corporation to take on more towns (from 23 to over 225), more service coverage, larger customer base and better service delivery. As a result, our revenues have greatly improved. This model has also enabled projects to set off timely as opposed to waiting for external funding approvals which could take a longer time period.

Management recognises that sustainable revenue management cannot be achieved without proper customer care in place. As pointed out before, management has put in place a customer relationship model (CRM) that collates information from the call centre and the billing system and relays to the field teams for action and eventual feedback. The CRM has a dedicated team to check customer response efficiency. Performance laggards are promptly identified and reprimanded. Regular customer surveys are also regularly conducted and information obtained used for effective customer relations management. There is a referral system (through email, digital media etc.) where customer complaints that have not been adequately handled at field team level are referred to headquarters by affected customers for redress. This has ensured a strong internal customer relations regulation mechanism.

4.7 RATIONALIZING FINANCIAL MANAGEMENT CONTROL

Strong financial management control is key in ensuring that finances are well managed for healthy cash flows, working capital and reduce risks of fraud and other forms of illicit activities. A number of value-adding reforms have been introduced by the new management to strengthen the management control environment. First, the internal audit department was restructured to separate post-audit from pre-audit (compliance). These two functions were previously merged, making it hard to check on each other. Currently, the post-audit department functionally reports to the Board while the compliance department reports to the Managing Director. This has resulted in a separation of roles and greatly improved corporate governance of finances of the corporation. The compliance department checks that all payments are effected in line with the approved policies and guidelines by the Board and Top Management.

The post-audit department carries out periodic audits of all operations, including finance management issues, seeks management responses and reports to the Audit Committee of the Board. The Audit Committee holds quarterly meetings in which management is asked to account, based on audit findings and recommendation. The performance contract with Government of Uganda also has an indicator on proportion of audit recommendations implemented by management and the resultant performance contributes to determination of incentives payable to management. Previously, the Audit Committee of the Board used to be merged with risk management. This was not good corporate governance. The new management and Board have created a separate Risk Management Committee, which has brought better focus on both functions.

The other aspect of finance management control was the strengthening of weekly cash-flow planning. The planning processes now incorporates implementation of priority milestones of the Strategic Direction. Weekly executive management meetings are also briefed and asked to have an input on selection of priorities of the weekly cash-flow to create common understanding and consensus. Off-track financial management trends (e.g., over-expenditure on consumption at the expense of development) are also brought to executive management meetings, especially in times of liquidity challenges, for discussion and realignment of priorities. This approach helps to keep all top managers engaged and supportive of key managerial decisions. Risk exposure events that are likely to affect good financial management practices are also regularly brought to executive management meetings for discussion and consensus on remediation measures.

These approaches have resulted into significant positive effects on good financial management and propensity to support attendant reforms. No wonder, since 2013 the corporation has continued to obtain clean external audit opinions, with no qualification of accounts. The speed of preparation of draft accounts for external audit has also significantly improved: by the first week of the first month after the financial year, draft financial accounts are normally ready. This is a big improvement compared to previous years, when draft accounts would be ready three months after the end of the financial year. As a result, currently, audited accounts are normally ready and approved by the Board of Directors within two months of the end of the financial year, compared to six months in previous years.

4.8 ENHANCING EFFICIENCY IN LOGISTICS MANAGEMENT

Good logistics planning and management is key if organisations' plans are to be effectively and efficiently implemented. In line with the Strategic Direction, there was need to strengthen the logistics management function of the corporation. There were three aspects of logistics management that needed transformational action: internal planning and coordination, collaboration with suppliers, and transportation to final destination of goods and services. The effects of not doing the latter efficiently were significant. There were internal administrative inefficiencies, poor relationships with suppliers and expensive delivery mechanisms to final destination.

A number of remediation measures were introduced to enable efficient logistics management. On the side of internal planning and coordination, the department of logistics was restructured, raising the profile of procurement and inventory management units. New positions with job descriptions and associated job enrichment opportunities were identified to create opportunities for career growth and hence improved staff motivation. This was in view of the past and future business growth envisaged in the Strategic Direction goals. As a result of restructuring, new job positions were filled with capable and motivated employees whose competences were compatible with anticipated job challenges. New procurement committees were constituted, following the provisions of the law. The selection of members of the committees was done in such a way that the competencies tallied with the immense challenges associated with reforms envisaged in the new Strategic Direction (honesty, competence, vision achievement and speed of action). An e-inventory system was also created internally by the corporation's IT department to improve logistics requisition management, and hence improved logistics management. This was in addition to the e-procurement system that had already been developed by the same department, as described earlier. Additional managerial incentives for logistics staff were introduced (both financial and emotional) to improve staff productivity and overtime work commitment. Security of goods and services in transit from suppliers to stores and to the final destination was also improved through recruitment of young and more committed security personnel. The procurement and inventory staff were encouraged to have periodic joint meetings, without interfering with the roles of each other, for better coordination. Capacity building of staff (for office and field teams) was also enhanced, through recruitment of a dedicated logistics training advisor. This was done following the expanded operations management mandate in the Strategic Direction.

Regarding collaboration with suppliers, periodic meetings were introduced. In those meetings logistics management targets are discussed with suppliers in order to ensure common alignment with the corporation's annual investment targets. The logistics management targets mainly relate to delivery time schedules, agreed quantities and payment of goods and services delivered. Performance gaps are also promptly identified and discussed for improvement. Security risks that can result into loss of goods and services in transit are also discussed and remedial measures put in place. Wherever necessary, capacity building of suppliers in view of changes in the law and specifications requirements is periodically carried out. Also discussed in such meetings are expected new products that may require modification of factory production lines in time to avoid costly logistics outages. Further, during such meetings complaints from users regarding technical and functional quality of services from the suppliers are also presented and discussed with a view of causing improvements.

On the issue of transportation, the huge amount of logistics movements that came with the expansion strategy in the Strategic Direction brought some challenges, initially. The corporation was used to annual pipe lengths of about 100 km. With a requirement of over 1000 km per year, this development brought huge transportation challenges. This was on top of transport for additional chemicals, plant parts and other operation and maintenance (O&M) materials. This challenge was managed through use of outsourced transport trucks where the field managers were asked to directly engage them at a local level. Materials would be sourced in bulk (to benefit from discounts) and transportation was decentralised for purposes of enhancing administrative efficiency and accountability. In addition, the corporation has acquired more trucks to cope with ever-increasing logistics transportation needs. The trucks have been mounted with a transport fleet management system to monitor vehicle movement and avoid possible abuse. As a result, goods and services are transported within two days from the date they are announced (by inventory staff) as ready for loading.

The logistics department also has a compliance unit that regularly checks and ensures fulfilment of procedures and guidelines. In most organisations the logistics department is the most prone to integrity issues. The compliance unit helps to form a third eye to guard against what can go wrong. The unit checks the physical movement of inventory goods and the correctness of documentation. Outliers are highlighted and corrected in a timely manner. Moreover, the corporation periodically registers fairly clean stock-taking exercises, even involving external auditors. The practice has been to employ people of the right behavioural competences in the compliance unit who measure up to the task.

Chapter 5

Customer and community perspective

The new Strategic Direction put particular emphasis on better customer service and community accountability. The essence of this step was clear: to ensure improved customer satisfaction with NWSC services and better collaboration with the wider community. Indeed, it is a myth for many publicly owned enterprises to appreciate the role of good customer care and public service accountability. But an orientation in this direction is central for any enterprise that aims to operate in a business-like manner. Its common logic that you cannot 'milk an unhappy cow'. The customers must be happy in order to guarantee high levels of willingness to pay for services. At the same time, a happy community offers a conducive environment for the utility to run its business without distraction. In the sections that follow, we explore some of the key focus areas and actions that the Strategic Direction considered and eventually implemented. Exhibit #29 highlights, broadly, what NWSC has gone through to strengthen its customer and community orientation.

Author: Silver Mugisha
doi: 10.2166/9781789060256_0107

EXHIBIT #29: FROM LENNY OTAI (2018), NWSC AREA MANAGER

Customer and community management has generally been enhanced through a number of ways. First, the environment has been positively enabled through increased autonomy for managers and staff in most business units. That has made it easy to introduce tailor-made stakeholder engagement activities which have improved customer/stakeholder buy-in. We have also had a visionary leadership at the top which has embraced change, innovation and looking at things from a different perspective (quite different from how everyone else sees or thinks) and working collaboratively to make things happen. There has also been a change of business model from a profit-oriented one to service-oriented kind of model which emphasises service above profits and surpluses. The exposure of the organisation employees locally and internationally – a lot has been learnt out there through benchmarking.

5.1 REINVIGORATING THE VISION, MISSION AND CORPORATE VALUES

The previous strategic plans of NWSC had a vision that aimed at turning the corporation into a leading water utility in the world. The area of emphasis of this vision was blurred and this tended to bring lack of clarity and focus in the protracted efforts of managers and staff. The new Strategic Direction and plans that followed sharpened the vision to bring out customer-centricity. Furthermore, the new Strategic Direction and corporate plans introduced a quality policy statement, for emphasis. The quality policy statement was characterised by provision of quality water and sewerage services to customers in collaboration with other stakeholders in an efficient and cost effective manner, while ensuring customer delight and continuous service improvement in an environmentally friendly manner. Exhibit #30 gives annotated strategies taken to enhance customer orientation and corresponding motivation among staff.

EXHIBIT #30: FROM MARGRET APOTO (2018), NWSC BRANCH MANAGER

I have keenly followed how NWSC has measured to its vision of moving towards being a leading customer centred utility in the world, its mission and values. Specifically, the zoning of the corporation has greatly improved service delivery to our customers. By zoning operations into Eastern Zone and Western zone (for Kampala); Central Zone, Western and South Western zone and North and North Eastern Zone within the corporation has greatly led to hands-on customer service. As a result, we have created delighted customers. An example of this is the awards NWSC has received in the recent past. The most recent is Peoples' Choice Award for the corporation's outstanding good customer care. The latter is a sign of a delighted customer base. There are many other awards we have received in the last four years. In addition, the corporation has created employment opportunities for so many Ugandans. This has been realized through rapid geographical expansion process by creating more branches to improve the coverage; more jobs have been created, as a result. Our movement towards achieving the vision has also been enabled by a motivated and engaged workforce – with enhanced pay and improved staff welfare. This has been evident in the prompt payment of staff salaries, bonuses and gratuity. We have also, occasionally, received rich, educative, motivating and encouraging messages from the Managing Director. In addition, staff promotions and empowerment has been achieved in the strategic plan. More training of staff within and abroad has also been witnessed. More so, scholarships have also been given to staff for further studies so as to have a learned/knowledgeable workforce. NWSC within the last four years has been able to construct a new home (head office) at Plot 3, Nakasero Road in Kampala. This has given a strong brand image to the corporation. With our Managing Director, NWSC have been able to move this far and achieve all the above because of a visionary, innovative, dynamic, and vibrant leadership.

Furthermore, past strategic plans put particular emphasis in the mission statement on efficient and effective provision of water and sewerage services using a host of managerial solutions to the delight of customers. With the new strategic direction, the mission statement underlined issues of sustainability, equitable and cost-effective service provision, good quality services to the delight of customers and environmental conservation. Clearly, therefore, signposts of a new approach to service offer were put in place. The corporate value system also went beyond previous tenets of professionalism, integrity and innovation. This time round, the Board and Management were determined to inculcate additional tenets of reliability, teamwork, excellence and results-orientation in the value system. This was aimed at building a strong foundation for requisite competences that were needed to fully anchor better implementation of new strategic plans.

In order to ensure meaningful community service, the new strategic plans also incorporated a theme that aimed at continuous improvement of services to communities for sustainable and equitable service delivery. Emphasis on community services was further enhanced by a catchphrase that hinged on water for all, for a delighted customer by a delighted workforce. The 'customer is king' mantra was also re-incorporated in the new strategic plans and effectively applied to drive cultural change in work ethics.

5.2 STRENGTHENING CUSTOMER INTERACTION MANAGEMENT

Customer interaction management is the process that companies follow to interface with its customers. NWSC has had a plethora of customer interaction management tools. Many of them had to be strengthened in order to match the customer care momentum envisaged in the new strategic plans. As already pointed out, the call centre facility had to be revamped. Due to the increased in the customer base over time, we decided in 2014 to expand the call centre facility, transferring it to a more spacious place with better working environment and configuration. Using knowledge benchmarked from best international practices in South Africa and the Netherlands, we bought new equipment, increased call agents (from 3 to 25) and introduced an out-bound facility for giving feedback to calling customers. Performance indicators and targets for call centre staff were also introduced to increase productivity. A system for quick relaying of concerns from customers to field operatives was also strengthened through the introduction of a computer-aided customer relationship module (CRM). The CRM also facilitates quick feedback to outbound staff and allows easy monitoring of actions by field managers. Performance gaps are easily identified and addressed by supervisors.

On top of the call centre, an active digital marketing team has been created to deal with external customers' concerns via various social media platforms. A culture of quick response and action on customer concerns on social media has been put in place. At the same time, the social media team has developed capacities to carry out digital intelligence so that bad and malicious public corporate publicity (via social media) can be proactively detected and managed. The social media platforms are also actively utilised to relay and communicate important information, plans and developments that have taken place in the corporation. This helped to market the NWSC brand nationally, regionally and internationally.

Another customer interaction management tool that has been transformed is the front desk office. Traditionally, the front desk is manned by staff without sufficient corporate information, yet they are supposed to assist key stakeholders that need vital information. This is not good for any enterprise that wants to operate in a business-like manner. Consequently, we had to raise the profile of front desk to a corporate relations management office. We deployed staff with good knowledge, skills and attitude to act as the first-impression face of the corporation. The front desk areas have also been refurbished and furnished to reflect expected levels of client service. The corporate relations staff at the front desk have been rebranded with motivating job descriptions and work gadgets so that everyone else feels compelled to work in that area.

Furthermore, we introduced periodic customer interaction sessions (barazas) where managers share achievements and challenges with customers in a transparent manner. The sessions are modulated by independent facilitators. Customers are allowed to share their experiences with managers and promises of improvement where necessary are made. The spirit is for these promises to become social contracts with customers and eventually help to impact design of future strategic and tactical plans for improved service delivery. In these sessions, political leaders are also invited to share their experiences and give guidance where necessary. Other customer interaction management tools include direct phone calls with managers, interactions on the corporation website and other public gatherings. Exhibit #31 illustrates what has been done to increase customer awareness.

EXHIBIT #31: FROM SHARON KARUNGI (2018), NWSC BRANCH MANAGER

Emphasis on customer engagement has helped in spreading the NWSC 'gospel'. Customers are now more aware of NWSC ongoing works and plans. This has also improved community acceptance and involvement in these works.

Figure 5.1 NWSC Management share achievements and challenges with customers during the Water Baraza at NWSC IREC Kampala. August 2014.

5.3 RE-SCOPING AND STRENGTHENING CUSTOMER SATISFACTION ANALYSIS

Customer surveys have increasingly become powerful tools of gauging the quality of utility offering, as a basis for continuous customer service. NWSC has had a long history of conducting customer surveys and analysis as a means of strengthening its customer and commercial orientation. However, up to 2013, all surveys and analyses were mainly being conducted internally and there were not sufficient independent checks and validation. In addition, customer surveys were only carried out annually with no intermediate tailor-made surveys to mitigate risks of customer service failure. For example, if there were lapses in the way new connections and disconnections for non-payment were carried out, there was no proactive mechanism of detecting service gaps. As a result, the corporation's customer service offering was mainly characterised by reactive approaches and concerns would go un-noticed for a long time without appropriate remediation measures. Exhibit #32 hints on ever-changing customer demands that organisations need to contend with for survival.

EXHIBIT #32: FROM FRANCIS BBALE (2018), NWSC SENIOR ENGINEER

NWSC has put great emphasis on customer satisfaction analysis. In the last four years, NWSC customers are no longer just 'kings' but 'overly demanding' and all staff have been tuned to providing a service that puts a smile on customers' faces. The responsive actions have been improved further with the establishment of a social media team (dealing with real-time engagement of customer complaints and ensuring prompt resolution and provision of reports and feedback alongside a fully ledged call centre department. The impact of these actions can be evidenced from awards that NWSC has bagged over the last four years in the field of effective communication.

These shortfalls were addressed in a number of ways. An independent customer service consultant was engaged to review the adequacy of the customer survey instrument that was being used internally and then carry out independent surveys to rebase the customer satisfaction index. Indeed, this exercise was carried out with minor modifications in the survey instrument and a new aggregated customer satisfaction index (CSI) of 78% (compared to an internally determined statistic of 85%) was assessed. Customer satisfaction index (CSI) is the figure obtained by dividing the average aggregated score of perceived customer ranking of customer service offering by the total possible score (on a Lickert Scale, e.g., 1–10 as used in this case), expressed as a percentage. The independently determined CSI then helped to act as a base for subsequent customer surveys. The slight drop in CSI was not, necessarily, representative of a drop in customer service but rather due to rectification of bias and other methodology-related issues. As a result, subsequent customer surveys adapted the modified survey instruments and survey methodologies, using internal staff but better trained.

Furthermore, intermediary surveys (most of them based on an 'as need arises' basis) were introduced to detect emerging lapses in customer service for immediate action. Some of the early warning signals have occasionally come from concerns on the social media, the call centre, informal interactions with customers and direct phone contacts with senior managers, and sometimes Board members. These types of surveys are carried out using tailor-made survey instruments by select groups of customer relations officers, based on random samples selected from affected customer groups. The findings and recommendations are shared with affected field managers, copied to senior managers (including the Managing Director) for immediate action. The field managers are normally asked to give timelines of expected remedial actions to be taken. This approach has helped to further strengthen customer service culture in the organisation, which has in turn improved the CSI to about 88% as of June 2017.

The culture of conducting customer surveys has also been introduced in the call centre operations for purposes of gauging service offering and behaviour by call centre agents. This has been deemed an important intervention, given that the bulk of customer interface operations take place in the call centre. If things go wrong in the call centre, the brand image of the corporation can be adversely affected. As part of this exercise, customers were also asked to comment on the timeliness of responses to customer concerns raised and if they give approval to quality of service. The findings are regularly shared with call centre agents with a purpose of continuous improvements. Identified good performance is rewarded, mainly through staff recognitions, promotions and job enrichment, as management deems fit.

5.4 INTRODUCING COMMUNITY COMMUNICATION PLATFORMS

As part of strengthening customer interaction, management also introduced a number of communication platforms to enhance customer engagement. The main rationale was to improve customer retention, advocacy and loyalty to the corporation. In order to achieve this objective, management looked at stakeholders with powerful constituencies through which the corporation would optimise its outreach.

5.4.1 Water Community Communication (WACOCO) clubs

The Water Community Communications (WACOCO) clubs were introduced to strengthen the bridge between the corporation and various community/civic stakeholders. These include: political leaders, religious leaders, civil servants, non-government organisations (NGOs) and judicial officers. The main objective of WACOCO clubs was to create a platform for sharing the goings-on, planned activities and challenges being faced by the corporation. At the same time, the clubs are used to get feedback and helpful information necessary for effective management of corporation operations. In other words, WACOCO clubs provide a good opportunity for collaborative management of the corporation for effective service delivery and accountability.

The clubs have a number of salient features that make them fully effective and integrated. They emphasise the requisite ecological cycle of water supply – environmental management, water supply and sanitation. At a WACOCO club stakeholder meeting, these three elements are emphasised and the community is sensitised about the interdependence of each aspect and how each can be sustainably managed for long-term service provision. During such meetings there is also an emphasis put on the elements that are necessary to ensure cost-effective operations and capital investments. These elements include: collaborative efforts to minimise non-revenue water, community support in minimising land compensation associated with community projects, community participation in providing cost-neutral labour for community projects, and smart partnerships in fighting illicit activities by field staff and possible community collaborators. WACOCO clubs are also used to share information on hygiene management through proper sanitation practices. In this regard, areas with limited penetration of water and sanitation services are shared and intervention plans agreed upon.

In order to make WACOCO clubs meaningfully interactive, incentive plans have been incorporated in the design. These include giving of awards/gifts to excelling community representatives. Such awards include bicycles, television sets, umbrellas, t-shirts, pens, corporation calendars and diaries. These appreciation moments have proved to be powerful give-back activities that create a strong sense of ownership, advocacy and loyalty. Of course, the incentive package is designed with a strong sense of cost-benefit orientation and balance. Consequently, the scope and frequency of awards is designed in such a way that there is seamless optimisation of costs and benefits. This is always done to minimise risks associated with similar programmes in the past where costs were incurred with no commensurate accruing benefits and/or return on capital employed.

5.4.2 Know-Your-Customer (KYC) approaches

The modern trend in customer service is that customer needs continue to change very fast and require resilient approaches by organisations to cope up. This need became more paramount with increased customer focus in the Strategic Direction. One powerful way of coping up with this challenge is employing a '*Know-Your-Customer*' (KYC) approach. In NWSC, this has involved attaching a dedicated staff member to each and every customer of the corporation. This was done by informing every customer of the telephone contact of the relationship officer. The telephone numbers are printed on customer invoices/bills for ease of reference. The concerned staff have been sensitized to ensure that they are available and ready to deliver required service when called upon. Delinquent behaviour, in this regard, is punishable according to the corporation's human resources manual and responsible staff are aware.

In order to ensure that concerned relationship staff do not find excuses relating to airtime on phones, the corporation negotiated a common-user-group (CUG) facility with telephone companies that enables full access to phones with limited off-air interruptions. This approach has further entrenched a customer service culture among staff. There are, of course, human-related challenges to this approach but at least it has raised the corporation's customer service frontier. There is now more awareness that staff need to be more available to serve customers since the corporation's motto is 'Customer is the Reason we Exist'. Service areas have also been restructured and allocated supervisors (normally designated as territorial leaders) to make sure that relationship officers are effective in their work.

5.4.3 Regular high-level stakeholder meetings

In Uganda (and indeed in all other countries), there are stakeholders that are in charge of very influential constituencies that are crucial for smooth operations. These include members of Parliament, judicial officers, elected political leaders at other levels, religious leaders, development partners, accounting officers of collaborating institutions, the media, non-governmental organisations (NGOs) and manufactures/industries. In order to ensure effective implementation of the Strategic Direction, we have had to keep a strong linkage with these stakeholders. Regular meetings are held to share achievements, challenges and future plans of the corporation. Efforts are made to sustain linkages between what the corporation is doing and possible reciprocal benefits therein for the stakeholders in question. Exhibit #33 alludes to measures taken to buy-in stakeholders to support organisational programmes.

EXHIBIT #33: FROM CATHERINE ILLEKOMORIT (2018), NWSC AREA MANAGER

I have noticed a major shift to stakeholder engagement in the last four years. There has been increased vigilance in customer and stakeholder engagement through several platforms and at all levels through programs like WACOCO and SWAS clubs. Vigilant marketing strategies at all levels, using several platforms have received appreciation from donors, government and other interested partners. This has also sold the corporate brand very highly and many partners are interested in supporting projects, either for knowledge sharing or to support their funding. Monitoring of strategic progress against the strategic direction through the continuous evaluation meetings and directorate meetings has also been a key success driver.

Sometimes, the corporation derives issues of discussion at meetings with these stakeholders, through regular corporate relations surveys. This is a pragmatic approach to issues identification. It is also a powerful tool to gauge authenticity of the corporation's offering and value proposition. The results of these stakeholder meetings are used to modify and improve the functional (behavioural) quality of the value proposition offered by the corporation. The frequency and scope of these meetings is majorly demand-driven, especially when the corporation wants to mobilise and enlist support for key policy directions like tariff adjustments, loan applications, capital budget support and organisational restructuring. Exhibit #34 further illustrates the importance of stakeholder mapping.

EXHIBIT #34: FROM IBRAHIM WALUSIMBI (2018), NWSC AREA MANAGER

We have witnessed serious stakeholder engagements in the last four years. Unprecedented stakeholder mapping and engagement which has seen the corporation tap into opportunities that we have not had before. SCAP100, winning a bottomless list of national and international accolades attest to this dimension.

Figure 5.2 Dr Eng Silver Mugisha (middle) poses for a photo with journalist and social media practitioners at NWSC Water Heroes stakeholders engagement meeting at NWSC-IREC Kampala. September 2016.

Figure 5.3 NWSC top management attending a partnership meeting with the Top Leadership of Uganda Police at NWSC IREC Kampala. June 2017.

5.4.4 Other ways of community penetration

There are other ways of community penetration that were introduced to improve public connectivity and accountability. The idea was to keep the costs of community engagement as low as possible and yet achieve maximum impact. These included meetings of key corporation staff, especially in regional offices, in areas of worship where people gather in large numbers and can easily be mobilised to listen to the communication. The other areas are parties and markets where also people of different types meet in large numbers. The common messages shared in such gatherings include tariff structures, new connection procedures, what to do to avoid disconnections for non-payment, collaborative reporting of leakages, the need for environmental conservation and the nexus with water supply.

Other means of community penetration include regular talk shows of area/regional managers about the service offering of the corporation. This has been decentralised so that the messages delivered are tailor-made and specific to problems affecting the communities in question. However, communication guidelines have been shared so that the messages delivered do not result in any backlash to the brand image of the corporation. As much as possible, managers and/or officers that are well informed about the bigger corporate picture and specific regional operations are allowed to speak for the corporation. In other words, we have created many quasi-public and communication officers to ensure a cost-efficient and effective way of engaging the public. Periodically, senior managers (including the Managing Director) also appear on regional media platforms to interact with listening public. A question and answer (Q&A) session is allowed in all cases to get accurate feedback for improvement in service offering.

The corporation's plans, achievements and challenges are also regularly shared through presentations to targeted audiences at national, regional and international level. These forums are particularly important because they are normally populated by people from different backgrounds who are keen to ask questions, critique strategies and give compliments where appropriate. This helps to give authenticity to the strategies being applied by the corporation and, at the same time, to create recipes for modifications. During such occasions, especially at conferences, the corporation participates in exhibition processes where many people are shown the current value propositions offered by the corporation as well as opportunities for partnerships where possible. Some of the communication tools used in such exhibition activities include: video clips, historic performance brochures, corporate plans, project briefs and consultancy services opportunities.

5.5 MANAGING DIPLOMACY

Managing diplomacy is an area that any organisation that wants to survive in a hostile environment and implement its strategic plans must seriously consider. From our experience at NWSC, the key constants in all this must be fair dealing, transparency and reciprocity. Optimising these parameters is quite a challenge that managers must face and handle in a frugal manner. Management of diplomacy must be looked at internally and externally with respect to the organisation. Exhibit #35 shows the fruits of inculcating diplomacy and customer care in effective management of stakeholders.

EXHIBIT #35: FROM APOLLO OMODING (2018), NWSC BRANCH MANAGER

Diplomacy also means that the organisation implements facets that create a positive work environment. Over the last four years, the corporation has done a number of things that have enhanced a positive work environment. There is a significant improvement in problem-solving processes through effective communication, departmental harmony and minimal bureaucracy. There is also a growing culture of focus on customer satisfaction and diplomacy, speed to task execution, individual responsibility to deliver results through team work, harmony and unity. Staff empowerment through continuous skilling, adequate provision of materials (facilitating work force with tools and equipment). The visible ambition of the board and top management to grow and transform the corporation is also a good sign for a growing organisation. All these initiatives could not be successfully carried out without due recourse to effective diplomacy with stakeholders.

During the last four years of implementation of the strategic direction at NWSC managing diplomacy has been an issue of focus. There have been three categories of stakeholders where delicate diplomacy management has continually been required: the staff and the Board of Directors on the internal side, and key decision makers (members of parliament, judiciary and other oversight bodies) externally. The key matters that normally call for delicate management of diplomacy are: recruitment of staff, corporate social responsibility requests and various forms of personal requests. Obviously, the equation underlying management of diplomacy in such matters can never be expected to be linear especially if the constraints of fair dealing, transparency and reciprocity come into play. At the same time, the three players (stakeholders) above require different approaches.

For internal staff, the approach has been to tell them that we all must share the vision of continuously improving the corporation, because it is 'the goose that lays the eggs'. Therefore, any personal requests made must respect principles of cost containment, belief from outsiders that we are doing things correctly, respect for the corporation's efforts to contribute to overall national development (given that Government is the sole shareholder), equity to all staff (in diversity) and most importantly, personal interests must never jeopardise the common interest of all staff. Of course, this calls for everyone to live and act as an example. With this frame of principles and all the time laying them on the table for staff with personal requests, the corporation has fairly built a culture of acting in the best interests of each other: put all-staff interests above personal interests.

The approach to the Board of Directors has been slightly different. The Chairman of the Board has taken leadership in telling fellow Board members that creation of a track record of good performance and leaving a legacy is paramount. This has been a song in every board meeting and this tended to create a sense of self-awareness and self-regulation among members of the Board. Hence the Board has been very supportive in managing diplomacy.

The approach to personal requests from external stakeholders has also been different. Over time, management has built a frame of fundamentals that have helped to mediate personal requests from external stakeholders and the need to maintain principles of fair dealing, transparency and reciprocity. This is where diplomacy comes into practical play. As management, we have emphasised the following fundamentals: we desire that all citizens should access water through our 'water for all' mantra; the accountability entities (inspectorate of government, public procurement and disposal authority, parliamentary accountability committees, the media, among others) are always watching to see correct things being done; we strive for equity (in diversity); and we shall apply high standards, as every Ugandan should want to see NWSC perform to international standards. By articulating these fundamentals, external stakeholders with personal requests are kept at some distance. As management, we also try to live by example. In this sense, we play management of diplomacy by principled opinion. It somehow works, with a few outlying opinions that require tailored approaches. What is critical here, is the recognition that management of diplomacy requires concerted application of emotional intelligence – endeavouring to make the other party believe in what you believe in.

5.6 THE ROLE OF CIVIL SOCIETY

Civil society is a key component in the governance of the water and sanitation in Uganda, including operations of NWSC. The civil society for water and sanitation in Uganda is organised under the Uganda Water and Sanitation NGO Network (UWASNET), which is the national umbrella for civil society responsible for effective delivery of water and sanitation services to citizens. UWASNET works directly with the Ministry of Water and Environment and liaises directly with key players in urban water, rural water, environment, forestry, climate-change, meteorology and water resources. UWASNET has a network of over 250 Non-Governmental Organisations that have been joining over time.

UWASNET is represented on most of the governance structures of the Ministry of Water and Environment including the sector working group, annual joint sector review meetings, joint technical review meetings, Water Policy Committee, top policy meetings and other ad-hoc committees formed as need arises. UWASNET prepares regular performance reports, which are presented to the Ministry of Water and Environment, providing scope for an independent evaluative framework. UWASNET also collaborates with the media and other arms of government like Parliament, Criminal Investigations Department of Police, Auditor General and Inspectorate of Government to enhance accountability and consumer protection.

5.7 REGULATION OF PUBLIC SERVICES OVER NWSC

Regulation of public services is increasingly becoming a novel concept especially where services are monopolistic in nature. Of course the perceived protection by a feeling of business monopoly, even for water and sanitation services, can only apply to mediocre managers. Even if there is limited room for product substitutability, there is definitely, room for functional competition. Managers in public water utilities face competition for: recognition by policy makers, preferential bill payment by customers, loyalty and scarce resource envelopes. It is only efficient public companies that stand to beat this type of new competition. Utilities that want to survive in turbulent reform eras have to work hard to beat these forms of competition.

However, some form of regulation is always necessary to reinforce managerial efforts to maximise value offering to citizens. For NWSC, the preferred form of regulation is through a performance contract with Government of Uganda, through the Ministry of Water and Environment. The three-year performance contract stipulates obligations, performance targets and investment requirements to be meet every year. It also has an incentive/penalty system that takes due consideration of performance trends. The contract is monitored by a performance review committee that works through the regulation department of the Ministry of Water and Environment. Other constituent elements of the regulatory framework include: Committees of Parliament, the Auditor General, the Solicitor General, the Procurement and Disposal Authority, the print and digital media, and the Parastatal Monitoring Unit of Ministry of Finance Planning and Economic Development. There is also the Uganda National Bureau of Standards and Ministry of Water and Environment Laboratory Section for water quality regulation. The regulatory framework of NWSC has proved to be fit for purpose, cost-efficient and acceptable to policy makers. It has ensured that there is continuous operational and financial performance improvement.

Chapter 6

Summary conclusions and recommendations

Increasing efficiency and effectiveness of water supply service delivery is a major challenge in closing the coverage gap and achieving universal access. This is, particularly, in the context of limited capacity and financial resources that most of the developing countries face. Therefore, development of innovative and cost-efficient ways of offering services at scale is needed to break new ground in delivering improved and sustainable water and sanitation solutions for all. Moreover, countries are required to create smarter and more resilient investments to cope with the impacts of climatic change.

Over the last fifteen years, the NWSC has designed and implemented service delivery improvement reforms that have shown promising results. The foregoing chapters presented narratives and exhibits that underpin key technologies, processes and measures that have, specifically, been implemented in the last four years. It is to be noted that in real-life situations, success is a road paved with failures and challenges. Therefore, what has happened in NWSC in the last four years can best be characterised as such. As presented in the previous chapters, many innovative things have taken place in the technical, financial, people and customer/community perspectives. This chapter presents ten top lessons learned and conclusions.

Author: Silver Mugisha
doi: 10.2166/9781789060256_0125

6.1 GENERAL LESSONS LEARNED

Every reform process comes with lessons learned. In this section we present ten top lessons that we can say we have observed and can be of importance to other utilities in developing (and to a large extent in the developed world) countries. It should be noted that there are many more lessons but the ones below stand out and underpin the key success factors at NWSC.

6.1.1 Strong focus on action

A certain Ugandan Bishop once opined to the author that… 'even from a theological point of view, faith without action is dead'. He insists that people should learn to act in accordance with what they believe: putting faith to action and vice versa. At NWSC, we have learned that what man can conceive and believe, he will one day achieve or receive. Nonetheless, there has to be action. Clearly, NWSC would not have achieved a drastic jump from a coverage of 23 towns to over 230 towns in a period of 4 years, if it was not a deliberate notion to take calculated risks and action – and moreover, doing all this while keeping cost and operational efficiency levels high. In this regard, employees have to be continuously reminded that management and leadership is an action and not a position. At NWSC, we have built a culture of speed of work, arguing employees to work 'as if there is a lion in the compound'. Consequently, every manager is evaluated based on the extent of practical outputs realised. This, however, cannot be achieved if the top leadership does not promulgate a culture of 'action speaks louder than words' and cascade to all levels of management.

At NWSC, we have also learned that balancing planning and action is a good performance driver. NWSC has a built a culture of always having a well-articulated vision and long-term strategy, operationalized through a plethora of tactical plans; all of these are formulated in a participatory manner involving the Board, management and selected external stakeholders. We have insisted that hypothetical planning yields sub-optimal results. As much as possible, key aspects of a service delivery plan should be pre-tested to gauge their practicality. It is also possible to include expert people that have seen the practical side of things before making them part of the plans. For example, one of the key construction methods that has helped to drastically increase new pipe coverage at NWSC has been direct labour sourced from communities. This was introduced and institutionalised by first pre-testing its viability with nearby communities in Kampala City. We learned that smart combinations of planning and action (and not one after the other) present better practical solutions to service delivery.

6.1.2 Putting an emphasis on service for all

It is hard for utility managers in developing countries to achieve universal coverage if they do not include strong focus on service for all people as part of their day-to-day emotional incentive structure. At NWSC, water for all, has been a catchphrase for all the strategic plans in the last four years. This has become a common 'song' in meetings, field activities, daily prayers and other stakeholder engagements. As a result, this has helped to influence budgeting, expenditure prioritisation, cash-flow planning and customer service practices. Everyone takes pride in how many more people have been served with good quality water services. The catchphrase has also strengthened orientation towards public accountability: citizens don't mind about internal processes so long as they get quality services. Those without services want to be given tangible and predictable promises that they will get them sooner than later. Because of this, NWSC has become more acceptable to the public and key stakeholders. Of course, promises have to be delivered, otherwise the utility loses public trust and confidence.

Service for all also means that the low-income (poor) population segments have to come into the planning and action horizon. This has implications for the tariff systems to be applied and implicit cross-subsidy mechanisms to keep services affordable but conservable. The management options for such population segments also have to be smartly tailored to local contexts. One has to look at income generation cycles, expenditure patterns and other behavioural issues pertaining to such communities. It is when the latter has been properly contextualised that utility managers can guarantee sustainable service provision in such communities.

6.1.3 The role of smart staff engagement

At NWSC we have learned that it is key that continuous efforts are made to create a positive work environment and the right conditions for all employees of the organisation to offer their best, committed to agreed targets. It is paramount to avoid instances of employee disengagement (or perceptions of unclear situations), where employees start guessing what the top leadership is up to. Keeping the employees aware of what is happening in the organisations (through a plethora of communication channels) and planned welfare schemes for them can be a strong motivator and a source of commitment to organisational vision and values. The task of keeping staff engaged must get the tone from the top leadership to build authenticity and acceptability. Deliberate efforts towards staff engagement by management at NWSC have helped to create a positive work environment. Most employees are fully attached to and enthusiastic about their work assignments and so take deliberate measures to promote organisational interests and the brand image.

6.1.4 Employee buy-in through incentive mechanisms

It is increasingly becoming good utility management practice to design innovative mechanisms of ensuring that those who do the work and achieve the results feel 'part and parcel' of efficiency gains and losses, if any. This is one way of creating a strong sense of ownership in what is happening in any organisation. NWSC has established a track record (over 15 years) of using smart incentive plans to catalyse performance by incorporating them in its strategic and tactical plans. In all cases, employees are rewarded or penalised for efficiency gains and losses respectively. Incentive plans are designed around key performance areas that require the utmost attention during the planning cycle. The plans are managed through a central monitoring and evaluation (M&E) unit that calculates and awards earned incentives on monthly basis, for operational units and annually for headquarters staff. Incentives for headquarter managers and staff are computed annually, based on an agreed incentive plan in the Government Performance Contract; the calculations are then reviewed and authenticated by the Regulation Unit of the Ministry of Water and Environment.

We learned that if incentives are to be earned, there should be no 'change of goal posts': they should be scheduled in the utility cash-flow plans and paid. That is when legitimacy is created among staff that good performance is rewarded, in turn improving commitment to work. We have also learned that incentive plans should be modifiable when performance improvement priorities change. However, changes must be negotiated in a transparent, fair and equitable manner.

6.1.5 Economic principles remain relevant

There is growing belief that water is a basic human right, which must be afforded to people at all costs. While this is true, the fact remains that utilities will continue to require funds to maintain, rework and expand systems. It, therefore, means that someone somewhere has to pay for the services. Actually, most times, providing services at far less than their cost of production means those who do not have are indirectly paying for those who do have, through a national taxation system. At NWSC, we have learned that tariff systems can be structured and differentiated to maintain a thread of fairness and affordability across all customer segments. Tariffs also need to be protected against erosion by external economic factors like foreign exchange risks, inflationary effects, power costs, among others. This process is currently being carried out through an indexation plan, which is applied in a transparent and objective manner.

What is critical, though, is that any tariff modifications must be well explained to citizens, putting an emphasis on the implicit value proposition improvement. At NWSC, we have learned that using comparisons of water tariff rates with costs of goods that are consumed on a regular basis by people can create a strong sense of acceptability and appreciation. For example, we have told our customers that the cost of 1 kg of beef meat is equivalent to about 120 and 400 twenty-litre jerrycans of NWSC water at domestic tariff rate and public stand post tariff rates, respectively. The cost of a half-litre of beer is equivalent to 40 and 120 twenty-litre jerrycans of NWSC water at domestic and public stand post tariff rates, respectively Matters of individual household resource allocation efficiencies and the need for homestead expenditure prioritisation must be elaborated and underlined.

6.1.6 Reinforcing collective leadership

Working as a team with one common purpose is key for a successful utility performance reform. At NWSC, we have learned that teamwork spirit must incorporate those in oversight responsibility, direct leadership action and operational level. A common understanding and unity of purpose must be established across all action levels without constraints of 'boss' elements. When proposals for innovative action are presented at whatever level, everyone must be motivated by agreed vision, goals, and value system of the organisation. There is a time in NWSC when those in oversight position would act with suspicion and reject most of the proposals brought to them by subordinates. This was extremely undesirable and did not help anyone in the end. If there are failures, leaders at all levels must see themselves through the mirror to apportion blame. When there are successes, credit should be passed over and cascaded to all participants. It is this approach that helps create a positive work environment in the organisation to encourage risk and initiative taking.

There is also another common leadership mistake that utility managers make. When they assume new positions of leadership, they choose to get rid of the team that has been running the show. At NWSC, we have learned that people need to be talked to and cajoled to adapt to new leadership expectations. It is, often, a painful exercise but the dividends are greater in the long run. Our experience actually shows that, with concerted change management efforts, people easily adapt to new leadership expectations and values. You only have to show them that you will serve without discrimination, nepotism, favouritism and cronyism and that you prefer the path of meritorious performance and thirst for productivity, irrespective of any form of sect. Companies that have followed the path of complete overhaul have paid an indirect price in terms of reduced (and sometimes delayed) work efficiency and productivity.

6.1.7 Leveraging information technology

It is a common fact that productivity growth is a multiplicative effect of technical efficiency growth and technical change. Therefore, utility managers have to invest a lot in leveraging information technology (IT) to cause technical change. This, combined with concerted efforts to increase efficiency will yield higher outputs. At NWSC, we have learned that extensive computerisation and modernisation of management systems makes work flow more efficient and quicker. We have achieved tremendous system modernisation in the area of financial management, revenue billing, customer bill payments, customer relations management, procurement and logistics management, pay roll compilation and employee leave management. Most of these systems have been designed internally, by the NWSC IT applications development staff, enabled through various incentive structures. More systems are being and will be developed, including prepaid meter solutions, document management systems and other capacity building modules.

6.1.8 Putting an emphasis on career paths and professional growth

There is a common misconception that employee career growth is only determined by level of knowledge, which is in turn a derivative of education level. Because of this, in the past, staff of NWSC have over-concentrated on going for further studies with no commensurate regard to other critical elements that underpin meaningful career growth.

These include (in addition to knowledge): skills (practice) and attitude. The latter is a function of a multitude of attributes including but not limited to emotional intelligence, team-working, positive orientation to work, continuous learning culture, networking and diplomacy. We have learned that enlightening staff about these key ingredients that buttress career growth helps to achieve it in a holistic manner. We have made it clear that attitude contributes to more than 80 percent of what is needed for an employee to succeed in work life and, therefore, it requires special attention. At NWSC, attitude among employees is improved through a host of coaching and mentoring activities at all levels of management.

We have also learned that advocacy for employee career growth is better entrenched when you link it to organisational business growth. Business growth at NWSC has been a function of network expansion, increased connection hook-ups, operational area (geographical) expansion, increase in water production and sales. All this calls for more labour in terms of more employees recruited and creation of high positions for promotion of managers and staff. Employees need to hear that the organisation is growing and that this trend is taking place in tandem with organisational restructuring which creates new positions to cope with increasing business scope. At NWSC, we have been encouraging employees to undertake holistic career growth making themselves ready to take up any new positions that arise, before we search externally. This message which gets cascaded to all levels of management has helped to create intrinsic motivation among employees who feel a strong sense of belonging to the organisation. During this process, we also emphasize that employees need to maintain a strong linkage between organisational goals (wants) and their individual expectations.

6.1.9 Enhancing individual and team accountability

Apart from use of financial and emotional incentives, we have also learned that carefully negotiated performance agreements (both at individual and group level) also help to enhance employee commitment to work. Such agreements must incorporate each party's performance obligations, smart targets, key assumptions and monitoring and evaluation system. At NWSC, the use of performance, autonomy, creativity and empowerment (PACE) contracts at team levels and individual performance agreements (IPAs) at individual levels have acted as catalytic instruments to performance improvement.

Having performance contracts in place is one part of the success equation. Periodic monitoring and evaluation of contract implementation is equally important. We have learned that monitoring needs to carefully examine processes and outputs. Process-oriented monitoring can easily be construed as micro-management that distorts credibility of accountability mechanisms. Therefore, excesses associated with these activities need to be avoided so that the operating units are optimally motivated to use their informational advantages to make things happen on the ground. The frequency of performance evaluations needs to be predictable and when such activities are being carried out, the principles of objectivity, fairness and transparency need to be observed to guarantee legitimacy and acceptability of results.

6.1.10 Emphasising external stakeholder engagement

Every utility manager must map out key external stakeholders and carry out effective engagement activities to achieve maximum support of reform activities. In the case of NWSC, the key external stakeholders include its customers, politicians, religious leaders, sister organisations, development partners, suppliers (service providers) and the media. We have learned that external stakeholders need to be viewed as partners in business: they should be given appropriate information that resonates well with their expectations. Regular stakeholder satisfaction activities through meetings and surveys need to be carried out. It is sometimes hard and costly (time-wise) to achieve full alignment with expectations, but it is a price each utility manager involved in performance reforms must bear. Where other activities that are not core mandates of the utility (such as source protection, sanitation improvement and flood control) have to be carried out, effective collaboration with relevant stakeholders is key.

6.2 SPECIFIC LESSONS LEARNED

Looking at what has happened in NWSC from 2013 to 2017, we see that the assumption of responsibility for much big spectrum of towns has several advantages. It brings more buy-in from political leadership, everyone feeling a strong sense of water service delivery to citizens. This in turn breeds support for new investment programmes due to increased perception of value for public funds. The NWSC case shows that rapid expansion must be matched with rapid capacity building of technicians and managers responsible for operating systems, through training and identification of right skills. Utility managers have to be innovative by coming up with fit-for-purpose organisational structures and performance-based pay systems that resonate with business growth. In the case of NWSC, the formulation of a performance-based structure that recognises the size of operational areas, geographical spread and business size was a great incentive for cost containment. Of course, this depends on the flexibility and delegation that the Board affords to management to be innovative and execute decisions.

It is important to inculcate a strong sense of ownership among staff and managers regarding changes to increase geographical expansion. At NWSC, there has been a continuous advocacy for the need to balance spending on recurrent needs and capital investment needs. The staff must understand and appreciate this strategic direction: slight delays in payment of staff allowances and gratuities while funds are channelled towards seed investments should be of no concern. Rapid expansion also means that managers rise to the opportunities presented by this trend: creation of opportunities of staff promotions, increased business volume and enhanced acceptability and support by local leaders. The proportion of headquarter staff to overall staff has reduced from 6% to 4%, reflecting economies of scale. As such, the scope of increased investment surpluses has increased. The downside to the rapid expansion relates to increased supervision burden and management control. Without a 'sharper' eye, the potential for things to wrong is high. However, with a strengthened internal audit system, NWSC central office has been able to maintain a strong oversight.

6.3 CONCLUDING OBSERVATIONS

Water utilities in developing countries will continue to suffer from ranking low on national budget priorities, even when water itself has been regarded as a basic human right by United Nations. It is clear that the water sector will not be getting out of funding challenges, soon. This trend is further exacerbated by perceptions of high tariffs, high unit capital investment costs (US$50–70 per capita in case of the urban sector in Uganda) and over-reliance on donor funding by sector practitioners. And yet, political pressure on utilities to increase service coverage keeps mounting. Even if resources to expand the network were available, with increasing water demand, the water production facilities would continue to become stretched, requiring resources for expansion and rehabilitation. The situation is not helped by extreme weather events (floods in the wet season and long spells of dry periods) which put pressure on water resources. The need for compatible climate-change resilience measures will continue to be vital. All these measures require sustainable sources of funding. Therefore, the need for effective and efficient management practices to a level where utilities can begin to generate a significant part of their capital investment requirements, internally, is sacrosanct. In order to achieve the latter, the narrative in the NWSC case study described in this book points to a number of conclusions.

A new public management model presents a good alternative for improving the performance of public water utilities. This is, however, conditional on strong leadership that characteristically places specific emphasis on customer care, strong commercial orientation, cost containment and resource allocation efficiency. In addition, specific attention has to be put on relevant stakeholder identification and engagement, infrastructure (business) growth and human capital development. In other words, a leadership orientation that optimises a people perspective and the production side of business is critical to organisational success. Furthermore, we need leaders who find a good mix of strategic planning and downstream tactical action plans that incorporate smart incentive arrangements to catalyse performance. Increasingly, use of well-designed performance contracts (at individual and team level) guarantees strong managerial commitment to achieve desired results.

A balanced score card (BSC) has proved to be a strong planning and implementation framework, which ensures that utility business is run in a holistic manner. The BSC tool is also a good stepping stone to planning a successful organisational restructuring activity. In addition, it helps utility leaders and managers to keep track of implementation of planned activities by staff within their jurisdiction and monitor outcomes (and associated risks) as a result of their actions. What is critical, is that utility leaders must focus on the strategic direction of the organisation and select a small priority data set to monitor, both financial and non-financial. After choosing the data to measure, baselines (normally based on historical performance) are set and facets to take corrective interventions, periodically, are put in place.

Most often, it is not the core mandate of water utilities to manage the environment. However, it is increasingly becoming critical that water utilities make deliberate efforts to collaborate on this important aspect to ensure sustainable raw water resources availability. Of course, it is possible to carry out investments to institute climate-change-resilient measures like ground water abstraction and raw water reservoir expansion. But these activities will still be heavily reliant on the finite nature of (availability) of water resources. Therefore, utility managers must collaborate with sister agencies to engage directly in raw water source protection measures such as tree planting, wetland restoration and other pollution control measures. The right species of trees must be planted in the right places to avoid escalating environmental problems instead of solving them. It is always better to plant environmentally friendly tree species that at the same time boost the economic activities of the communities. It is also possible for utility leaders to look for partners in the business of environmental protection who consider this activity as an incentive to boosting their brand image. In Uganda, for example, Parliament and schools are proud to engage in environmental protection activities as part of their corporate social responsibility (CSR). There are also civil society organisations (CSOs) and community-based organisations (CBOs) that are purposed to advocate for environmental protection. These can be good partners to work with.

In addition, there is a common belief that utility management should stay clear of politics of the country. This orientation is far from the reality of how things happen. It is actually close to suicide for public water utility leaders to attempt to stay aloof from the politics of the day. This is partly because most important decisions affecting proper running and reforms of water companies are taken by politicians, ostensibly on behalf of the people they represent. Therefore, staying away from politicians means that there is a fertile ground for creating informational asymmetries, which can in turn result into uninformed decisions. Moreover, if public water utilities want to stand a strong test of public accountability, why should their leaders shy away from engaging peoples' representatives, the attendant price of doing it, notwithstanding? There are many ways politicians can be meaningfully engaged in creating a semblance of common understanding, while at the same time avoiding groupthink. There are no right prescriptions regarding how politicians (and similar minded stakeholders) can be meaningfully engaged. It is simply a decision a utility leader takes and often becomes positive about. It is an iterative and risky undertaking that requires good use of emotional intelligence skills and, eventually, improved through experience by doing.

The other important consideration for sustainable utility management is to put an emphasis on 'do it yourself' orientation. If a significant number of utility managers (starting from the top) are motivated by 'we can it ourselves' attitude, they will exert maximum efforts in generating ground-breaking turnaround ideas that can cause a difference. This will minimise the tendency to always sit and wait for someone to help: mainly through extensive use of costly consultants. The problem is exacerbated by the approach external consultants take: tell me what you want and I do it. How will a utility manager who does not know what should be done tell a consultant what to do? Clearly, this is double jeopardy. The NWSC case study has shown that, with a strong 'I can do it myself' approach, productivity improvement has taken place and continues to do so, failure risks notwithstanding. Of course this approach requires continuous capacity building through benchmarking, research and training. But the outcomes are more sustainable, predictable and less costly in the long run.

The other important input for efficient utility management is strong leadership. Public water utilities need leaders that value their workers (followers). They should champion cost-effective staff welfare systems, demonstrate transparency, fair dealing and empathy. They must also put particular emphasis on continuous staff development so that staff get a strong sense to support performance reform activities. Of course, this requires continuous learning through acquaintance with new trends in leadership, benchmarking and training. Increasingly, trends in successful leadership practices are emerging through research into experiences and practices in successful companies: particularly those that have moved from good to great (like General Electric, Apple and Google). It can be helpful to adapt some of the good leadership techniques from such companies to a local context.

The Sustainable Development Goals (SDGs) provide unprecedented political will to address water challenges in low-, middle- and high-income countries. The biggest opportunities lie in the developing countries, where population growth and rapid urbanisation means that the demand for water and sanitation services is greater than ever before. The NWSC case study illustrates steps that can move a utility towards the frontier of water service for all. The SDGs, and in particular SDG6, supposedly, provide a mechanism for mobilising political and financial support for delivering the water and sanitation infrastructure required to achieve them. It will be an issue of equity benefiting the one who is vigilant. It cannot just be running around, but also fronting transparency and a good track record of implemented plans, while at the same time giving the rationale behind project and programme proposals. If there is a significant chasm between presented objectives and policy makers' strategic expectations, there will be problems of acceptability and alignment. Utility managers will therefore have to maintain interoperability with other national development plan strategic objectives. The SDGs require much more than incremental improvements. They require a new mind-set in order to give rapid and sustainable access to water and sanitation services for everyone in a way that reduces inequity in cities and communities.

In the last fifteen years, NWSC through its External Services Department, has played a central role in bridging the gap between public and private-like service mentalities of water utilities in several developing countries. Working in a partnering manner, NWSC has helped to catalyse and diffuse a proliferation of performance reforms in several countries. It is possible for utility managers in developing countries to develop the right mind-sets and leadership development practices that yield sustainable water managerial solutions. There is need to develop a strong sense of self-confidence and can-do-it-yourself orientation rather than over-reliance on external support to provide expertise and financial resources. There is nothing holding water utilities back. It is a matter of determination, right attitude, doing what it takes and having the right resilience thresholds. The sanctity of the nexus between sustainable success and being comfortable with disruptive practices that catalyse requisite innovation and knowledge should not be underestimated.

Index

IWA Publishing's authorised EU representative for General Product Safety Regulations is Diane D'Arras, 15 rue Duret, 75116 Paris, France, e-mail: safety@iwap.co.uk.

Printed and bound by CPI Group (UK) Ltd, Croydon, CR0 4YY
12/05/2026
02108880-0007